M-Commerce
Technologies, Services, and Business Models

M-Commerce
Technologies, Services, and Business Models

Norman Sadeh

Wiley Computer Publishing

John Wiley & Sons, Inc.

Publisher: Robert Ipsen
Editor: Carol A. Long
Assistant Editor: Adaobi Obi
Managing Editor: Pamela Hanley
Composition: Benchmark Productions Inc., Boston

Designations used by companies to distinguish their products are often claimed as trademarks. In all instances where John Wiley & Sons, Inc., is aware of a claim, the product names appear in initial capital or ALL CAPITAL LETTERS. Readers, however, should contact the appropriate companies for more complete information regarding trademarks and registration.

This book is printed on acid-free paper. ♾

Published by John Wiley & Sons, Inc.

Published simultaneously in Canada.

This publication is designed to provide accurate and authoritative information in regard to the subject matter covered. It is sold with the understanding that the publisher is not engaged in professional services. If professional advice or other expert assistance is required, the services of a competent professional person should be sought.

Library of Congress Cataloging-in-Publication Data:

Sadeh, Norman.
   M-commerce : technologies, services, and business models / Norman Sadeh
      p. cm.
   "Wiley computer publishing."
   ISBN 0–471–13585-2 (paperback : acid-free paper)
   1. Electronic commerce. 2. Cellular telephone equipment industry.
3. Mobile communication systems. 4. Wireless communication systems.
5. Internet. I. Title.
HF5548.32.S16 2002
   658.8'4–dc21   2002000305

To my wife, Patricia

CONTENTS

ACKNOWLEDGMENTS

This book is an attempt to share what I have learned about mobile commerce over the past several years. Much of my experience in this area is a direct result of my activities at the European Commission, where I served as chief scientist of the European research initiative in "New Methods of Work and Electronic Commerce." There my colleagues and I were responsible for launching several hundred R&D projects in wired e-commerce, mobile commerce, and related areas. With each project typically involving five to ten organizations from industry and academia, it would be impossible to thank each of the many individuals and companies that, in one form or another, have contributed to my understanding of mobile commerce. I would like however to reserve some special thanks to Rosalie Zobel, director of the initiative in "New Methods of Work and Electronic Commerce," her predecessor, David Talbot, and my many collaborators at the Commission for giving me a chance to be part of a program that has played a key role in bringing Europe to the forefront of the Internet revolution.

More recently, since my return to the United States, I have had a chance to gain a deeper understanding of the mobile commerce scene in Asia and North America through my teaching and research activities in the School of Computer Science and the Institute for e-Commerce at Carnegie Mellon University (CMU), as well as through several consulting engagements. I would like, in particular, to thank Raj Reddy, head of the Institute for Software Research, International (ISRI), for his support and for giving me the time to work on this book while restarting my academic activities. Special thanks also to Michael Shamos and Tridas Mukhopadhyay for inviting me to join the Institute for e-Commerce. This book would not have been possible without the feedback from and many discussions with executives and students attending my lectures at CMU. I would like to thank Jay Gohil (Nokia), Enoch Chan (now with CMU), and Takeshi Saito (Corporate Directions Inc.) for bringing particularly useful material to my attention.

I would also like to acknowledge the important role of Johan Hjelm (Ericsson) who not only encouraged me to write this book but also provided me with invaluable feedback at different stages of the writing process.

The editorial team at Wiley has been outstanding. Special thanks to Carol Long, my publisher, for not giving up on me after I failed to meet some early deadlines and to the other members of the team, Pamela Hanley and Adaobi Obi Tulton.

Last but not least, I would like to thank my wife, Patricia, and our respective families for putting up with my strenuous schedule over the past several months. While, unfortunately, this lifestyle is nothing new to them, this book at least gives me the opportunity to thank them for their love and unfailing support.

Norman Sadeh
Pittsburgh, PA
sadeh@cs.cmu.edu

The M-Commerce Revolution

M-Commerce: What's the Buzz All About?

Introduction

The face of the Internet is changing. Within just a few years, more people will be accessing the Internet from mobile phones, personal digital assistants (PDAs), pagers, wristwatches, and a variety of information appliances other than desktop PCs. We are quickly approaching the mark of 1 billion mobile phone users and, while today only a fraction of existing mobile phones are Internet-enabled, the situation is fast changing. This is especially true in countries such as Japan and Korea, and in Europe. However, other countries such as the United States are not far behind. In early 1999, NTTDoCoMo launched its i-Mode mobile Internet portal. Within less than three years, the service had grown to 30 million users, in the process generating several billions of dollars in revenue in the form of subscription fees and increased traffic.

Mobile commerce, or *m-commerce*, is about the explosion of applications and services that will become accessible from Internet-enabled mobile devices. It involves new technologies, services, and business models. It is quite different from *traditional* e-commerce. Mobile phones or PDAs impose very different constraints than desktop computers do. However, they also open the door to a slew of new applications and services. They follow you wherever you go, making it possible to access the Internet while walking down the street with friends and family, or while driving, looking for a nearby restaurant or gas station. As the Internet finds its way into our purses or shirt pockets, the devices we use to access it are becoming more personal too. Already today, mobile phones and PDAs know the telephone numbers of our friends and colleagues. They are starting to track our location. Tomorrow, they will replace our wallets and credit cards. One day, they might very well turn into intelligent assistants capable of anticipating many of our wishes and needs, such as automatically arranging for taxis to come and pick us up after business meetings, or providing us with summaries of relevant news and messages left by colleagues. However, for all these changes to happen, key issues of interoperability, usability security, and privacy still need to be addressed. With hundreds of billions of dollars at stake, industry is working furiously to turn this vision into reality.

In writing this book, my objective is to introduce you to the technologies, services, and emerging business models associated with m-commerce, and in the process, shed some light on ongoing developments in this fast-moving and often confusing area. The material I present is based on numerous interactions I have had in recent years with a wide range of mobile Internet players, from handset manufacturers, to telecom operators, banks, Internet portals, Wireless ASPs, service providers, industrial forums, and standardization bodies to name just a few. Many of these interactions took place while I was serving as chief scientist of the European R&D initiative in "New Methods of Work and Electronic Commerce" at the European Commission. More recently, with my return to Carnegie Mellon University in 2001, I have also had a chance to gain a closer understanding of ongoing developments in the United States and Asia through various research, executive education, and consulting activities. In this book, I attempt to share with you some of what I have learned. The text is organized around eight chapters and follows the same flow as the mobile commerce courses I teach at Carnegie Mellon University to executives and e-commerce master's students.

In this chapter, we attempt to define m-commerce. We review the forces that underpin the "m-commerce revolution," looking at the proliferation of mobile devices, the emergence of the mobile Internet, and the transition to third-generation mobile communication technologies (3G). You will also get a first glimpse of the many new applications and services that are the foundation of m-commerce. This includes an initial discussion of the new usage scenarios and business models of m-commerce. In the process, you will gain a better understanding of what differentiates m-commerce from "traditional" e-commerce. To conclude, a summary of each chapter is provided along with a guide for how to read this book depending on your specific background and interests. As you will see, we have tried to keep the text modular to accommodate as broad a range of readers as possible. We hope you will find it convenient to navigate. Enjoy the reading!

What Is M-Commerce?

According to Durlacher Research, m-commerce is defined as *"any transaction with a monetary value that is conducted via a mobile telecommunications network"* (Durlacher 2000). A somewhat looser approach would be to characterize m-commerce as the emerging set of applications and services people can access from their Internet-enabled mobile devices. As this definition suggests, there are many dimensions to mobile commerce. Rather than attempting to come up with a cryptic one-size-fits-all definition, a more practical approach is to look at some of the many forms of mobile commerce found today.

NTTDoCoMo's i-Mode Portal

By far the most successful and most comprehensive example of m-commerce today, i-Mode is the mobile Internet service launched in early 1999 by Japan's NTT DoCoMo. As of early 2002, the service just reached the 30-million user mark, a resounding success in a country of 126 million people. The service, which is available for a monthly fee of about $3, offers a broad range of Internet services, including email, transaction services such as ticket reservations, banking, shopping, infotainment services (for example, weather, sports, games, and so forth) and directory services (see Figure 1.1). Most of these are provided by third-party content providers.

The early success of i-Mode can be attributed to a number of factors. Some would like to explain the rapid adoption of the service as a simple reflection of the low number of PCs and wired Internet users in Japan. Japanese people, they say, saw i-Mode as a cheap way to access the Internet and have email. However, while email is by far the most successful i-Mode application today, there is much more to the success of DoCoMo's mobile Internet portal. The deployment of i-Mode over

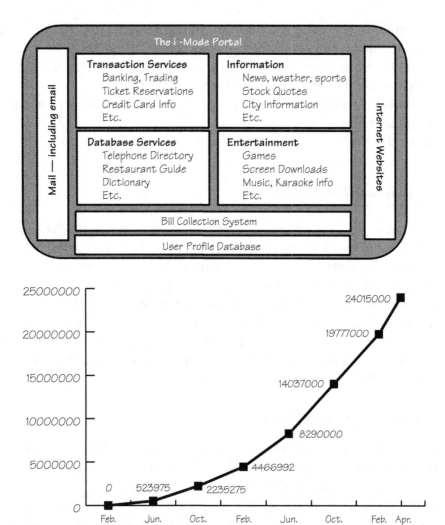

Figure 1.1 The i-Mode success story: Access to a plethora of mobile Internet services generates exponential growth.

DoCoMo's PDC-P network, a packet-switched technology that offers *always-on* functionality, making it possible for users to keep their devices on while only paying for actual traffic, is certainly one important factor. However, beyond this, i-Mode would never have been able to sign up as many subscribers if it was not for the critical mass of content providers it managed to assemble from day one.

The i-Mode programming environment was originally based on Compact HTML (cHTML), a subset of HTML—we will see in Chapter 4 that DoCoMo is now moving to XHTML. Being closer to the language of the traditional Internet than early versions of WAP, which relied on the WML language, cHTML made it much easier for content providers to develop i-Mode compatible services. In addition, DoCoMo extended the language specification to include convenient tags, such as an automatic dial tag that allows users to initiate telephone calls by clicking on a link. For example, when accessing a restaurant Web site, the user can click to call the restaurant and book a table, without having to disconnect from i-Mode.

Today, i-Mode can boast well over 1,000 official content providers, and tens of thousands of unofficial ones—*volunteer sites*, as they are referred to by DoCoMo. Official sites are accessible directly through the i-Mode menu, while unofficial sites require typing a URL. Applications for official status are reviewed by DoCoMo based on quality, usefulness, and accessibility of Web sites. Because they are accessible through the i-Mode menu, official sites tend to see much more traffic than unofficial ones. DoCoMo also uses its customer profile database to customize its portal to the needs and interests of each individual user. As Figure 1.2 shows, the most popular category of services, email aside, is entertainment, which as of year end 2000 accounted for 64 percent of i-Mode transactions. The popularity of these services also reflects the young age of many i-Mode users.

Another key element of DoCoMo's success has been its ability to turn the popularity of its services into a profitable business. In contrast to the wired Internet, where many popular services are still struggling to come up with a viable business model, DoCoMo relies on *multiple sources of revenue*, one of the keys to mobile commerce profitability. First, it charges a flat monthly fee for Internet access, around $3/month. In addition to also charging for traffic, DoCoMo offers official sites to collect monthly subscription fees on their behalf (see Figure 3.3). This involves collecting the fees from subscribers through DoCoMo's regular billing cycle, and ensuring that only authorized users are allowed to access these for-a-fee services. This is enforced by DoCoMo's i-Mode servers, through which all requests have to flow. In return, DoCoMo keeps a small percentage of the subscription fees it collects on behalf of content providers—currently, 9 percent. This arrangement makes it significantly easier for content providers to charge for their services without having to worry about setting up complex billing systems. Within one year of its

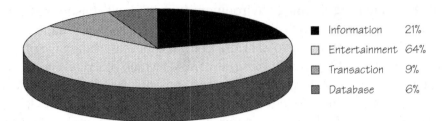

■ Information	21%
☐ Entertainment	64%
▨ Transaction	9%
▩ Database	6%

Figure 1.2 Among m-commerce services, entertainment dominates, reflecting the young age of the i-Mode user population.

Source: Infocom Research, *Second i-Mode User Survey*, September 2000.

start, i-Mode was already credited for an annual $120 ARPU (Average Revenue Per User) increase, $30 in the form of subscription fees and $90 in the form of increased traffic. By late 2000, that figure had doubled, with increased data traffic revenue of 2100 Yen per month per user— around $200 per year.

In summary, while some pundits still dismiss the i-Mode story as a purely regional phenomenon reflecting the peculiarities of Japan's Internet scene and DoCoMo's dominance of the Japanese wireless market, this view ignores a number of key factors. In truth, i-Mode shows us what it takes to succeed in the m-commerce arena:

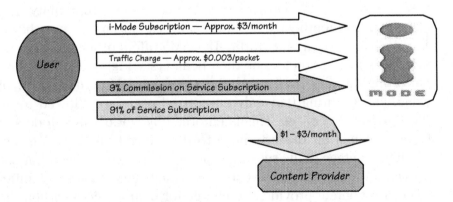

Figure 1.3 The i-Mode business model.

- Ease of use.

- Interoperability.

- A solid business model and a plethora of compelling content through no-nonsense partnership arrangements with a critical mass of third-party providers.

Today, DoCoMo is building on its success at home and looking for ways to replicate it abroad through partnerships with AOL Time Warner and key mobile telecom operators in the United States, Europe, and elsewhere in Asia. This includes minority-stake investments in companies such as U.S. operator AT&T Wireless, Dutch operator KPN, Hong Kong's Hutchison, Taiwan's KG Telecom, and more recently, Korea's SK Telecom.

Nordea's WAP Solo Mobile Banking Service

Japan is not the only place where mobile commerce has been making rapid headway. Nordic countries, home to almighty Nokia and Ericsson, have also seen their share of early successes. Nordea, a long-time pioneer in Internet banking, was the first to offer its customers WAP banking services through the launch of its WAP Solo portal in October 1999.

The portal makes it possible for customers to pay bills, check balances, review statements, or trade shares from their mobile phones (see Figure 1.4). It is fully integrated with Solo's other online banking channels, which include wired Internet access and Internet-enabled cable TV access. This integration ensures that WAP Solo customers are presented with a consistent experience, independently of the access channel they select—same look and feel, and same logical set of menus.

Another significant feature of WAP Solo is its shopping mall, "Solo Market," which gives customers access to over 600 merchants. Customers can browse merchant sites from their WAP-enabled phones. When they decide to make a purchase such as a movie ticket or a bouquet (see Figure 1.5), they are transferred back to the WAP Solo site, where they can securely pay for their purchases while selecting from a number of payment options—not just from Nordea bank accounts, but also from third-party credit cards. Nordea charges a monthly fee of 4 Finnish Mark—about $0.60 US—for access to its WAP Solo Market.

In less than two years, WAP Solo has garnered a substantial portion of Nordea's 2 million online banking users. Like DoCoMo's i-Mode—which,

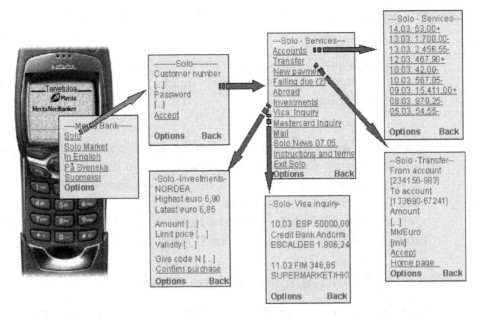

Figure 1.4 Nordea's WAP Solo banking portal: customers can check their statements, pay bills, wire money, trade shares, or make purchases from their mobile phones.

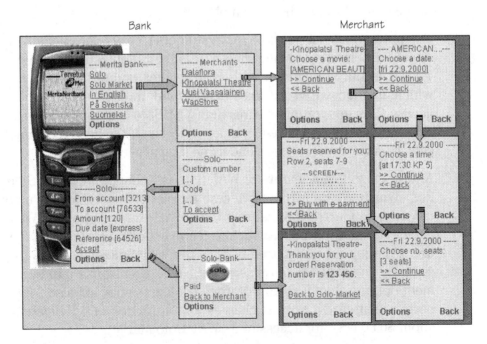

Figure 1.5 Purchasing movie tickets with WAP Solo.

by the way, also provides mobile banking services—WAP Solo's success can be attributed to:

- Its ease of use, which includes its integration with other online channels.
- The convenience of a site from which customers can access a critical mass of merchants and select from a broad set of payment options.

Nordea's business model, however, is quite different from that of DoCoMo's. It is part of a broader strategy of building a close and long-lasting relationship with its many clients, both private consumers and merchants, through as comprehensive a set of services as possible. As such, WAP Solo can also be viewed as a response to the potential threat of being dis-intermediated by mobile telecom operators, which, like DoCoMo, are eager to leverage their billing relationship with customers and grab a slice of every m-commerce transaction.

By offering a credible payment alternative, services such as WAP Solo make it possible for banks and credit card issuers to compete with mobile telecom operators for a slice of the m-commerce revenue pie. We discuss this further in Chapter 2, "A First Look at the Broader M-Commerce Value Chain." Nordea is also a founding member of the Mobey Forum, a broad industrial consortium whose objective is to open the m-commerce value chain through the promotion of open standards for mobile financial services. We further discuss the activities of Mobey and related standardization initiatives such as the Mobile Electronic Transaction (MeT) initiative or the Mobile Payment Forum in Chapter 5, "Mobile Security and Payment," and provide an extensive discussion of WAP in Chapter 4, "The Mobile Internet."

Webraska's SmartZone Platform

Mobile commerce is not just about large telecom operators and banks. It is also about a myriad of startups, each trying to conquer a particular segment of the m-commerce value chain. They include content providers as well as a new breed of companies commonly referred to as Wireless Application Service Providers (WASPs) that focus on the provisioning of applications and services to other players across the value chain. Like their wired counterparts, WASPs can take many shapes and forms, from providing packaged applications such as games or directories of services to hosting mobile portals or providing wireless access to enterprise applications.

One such application is SmartZone, commercialized today by French startup Webraska following its merger with Silicon Valley startup Air-Flash in September 2001. SmartZone focuses on what many have come to view as one of the most promising segments of the mobile commerce market, namely the development of location-sensitive services. Rather than having its own portal or directly offering its services to mobile users, Webraska sells its SmartZone platform to mobile portals and wireless carriers, who in turn customize pre-packaged location-based services for their mobile users. Using the resulting location-sensitive services, people can look for directions to nearby restaurants, ATMs or gas stations. They can find out about movies playing in their area and reserve tickets. A package is also provided to facilitate last minute activity planning with groups of friends such as deciding on a place where to have dinner.

Examples of mobile portals and telecom operators using SmartZone include AT&T Wireless and its PocketNet mobile portal, the Orange service in the UK, Germany's Viag Interkom and E-Plus Mobilfunk services, to name just a few. Webraska typically charges its clients a fixed monthly fee along with an incremental usage fee, which varies from one agreement to another. The company also offers revenue sharing for advertisers. In fact, as Figure 1.6 illustrates, Webraska itself relies on a number of partners to deliver its services. This includes alliances with location tracking providers such as Cell Loc or CPS. Most of its content also comes from partnerships with companies such as Last Minute.com for plane tickets, WorldRes.com for hotel reservations, or Zagat Survey for restaurant ratings. If this all seems rather complex, welcome to the world of m-commerce, where business partnership is the name of the game.

There are a number of factors contributing to the demand for mobile location-sensitive services. First, there is convenience. The limited screen size and low data rates associated with mobile Internet devices as well as the time critical nature of many of the tasks mobile users engage in, such as looking for directions, make it particularly important to have services that can provide location-relevant content. Another important factor is the rapid deployment of accurate location-tracking technology that will make it possible to pinpoint the location of mobile users, saving them the tedious task of having to manually enter this

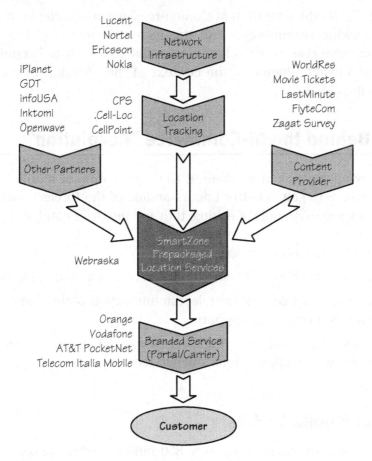

Lucent
Nortel
Ericsson
Nokia
Network Infrastructure

iPlanet
GDT
infoUSA
Inktomi
Openwave

CPS
.Cell-Loc
CellPoint
Location Tracking

WorldRes
Movie Tickets
LastMinute
FlyteCom
Zagat Survey

Other Partners

Content Provider

Webraska

SmartZone Prepackaged Location Services

Orange
Vodafone
AT&T PocketNet
Telecom Italia Mobile

Branded Service (Portal/Carrier)

Customer

Figure 1.6 Webraska's SmartZone value chain: A simplified view.

information. For instance, mobile carriers in the United States are progressively required to be able to pinpoint the location of users dialing the 911 emergency number—originally, phase II of the Wireless Enhanced 911 regulation, as it is commonly referred to, required that, by October 2001, mobile operators be able to pinpoint the location of 911 callers within 125 meters 67 percent of the time, though this deadline was recently relaxed by the United States Federal Communications Commission (FCC). Surveys suggest that users are willing to pay extra for the convenience of location-sensitive services that take advantage of such functionality. The end result is a sector that analysts predict could generate upwards of $10 billion a year within the next few years.

In Chapter 7, "Next-Generation M-Commerce," we take a closer look at location-tracking technologies and the many services and business models they give rise to. The chapter also discusses related standardization activities conducted in the context of the Location Interoperability Forum (LIF).

The Forces Behind the M-Commerce "Revolution"

Now that we have looked at some of the early forms of mobile commerce, it is time to gain a better understanding of the forces underpinning its emergence. Broadly speaking, they fall into four categories:

- Proliferation of mobile devices.
- Convergence of mobile telecommunication networks and the Internet.
- Transition to third-generation telecommunication technologies and the higher data rates they support.
- Emergence of a broad set of highly personalized, location-sensitive, and context-aware applications and services.

Proliferation of Mobile Devices

By the end of 2001, there were over 850 million mobile phone users worldwide—or about 14 percent of the world population—with annual sales of over 400 million telephones worldwide. Over the past decade, mobile phones have evolved from devices reserved to a select few to mass-market necessities and fashion accessories. In cities in Japan, Korea, Sweden, or Finland, it is nearly inconceivable for someone not to have a mobile phone, as the devices have become such an integral part of daily life (see Figure 1.7).

One area that has seen the most spectacular growth in mobile phone ownership is Europe, thanks to deregulation of the telecom industry and the adoption of a single standard, Global System for Mobile Communications (GSM), making it possible for users to roam across the entire continent. Today, GSM is the most popular mobile communication standard with two thirds of all mobile phones.

As markets in Europe, Japan, and North America slowly reach saturation, growth in mobile phone ownership will be driven by demand in

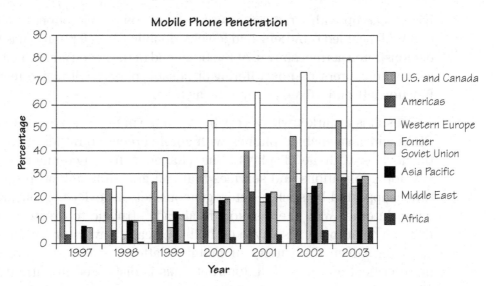

Figure 1.7 Mobile phone penetration by region.

Source: EMC Database.

Asia. In July 2001, the number of mobile phone users in China had already reached the 120-million mark—just higher than in the United States at that same time—and was growing at a rate of nearly 1 million per week. In light of the fact that China's population is 1 billion plus (close to 1.3 billion at the time this chapter was written) and increasing at one birth per second, the prospects for continued growth look fairly good.

While telephones dominate the mobile scene, other devices such as PDAs and pagers will also contribute to the growth of m-commerce. In 2000, customers bought over 6 million PDAs worldwide, with that figure expected to exceed 20 million in 2002. Mobile phones are also starting to display many of the same features as PDAs. Many sport larger screens and support personal information management (PIM) applications such as calendars and contact lists. In fact, a number of device manufacturers have started to offer true hybrids such as Nokia's 9210 Communicator—a mobile phone and clamshell PDA all in one—Handspring's VisorPhone, or Motorola's Accompli 008. High-end mobile phones are also becoming increasingly powerful, with the introduction in 2001 of Java-enabled models. Mobile handsets are also merging with other consumer electronics devices such as cameras or MP3 players. Over time, some of these devices will become small enough to be wearable. Already, Samsung and other manufacturers have come up with wristwatches that double as mobile phones, pagers, and WAP-enabled browsers, and jewelry doubling as cell phones has been demonstrated at tradeshows. At the other end of the spectrum, some handset manufacturers are now offering disposable phones with fairly minimal functionality and a fixed amount of call time.

The result is a market with a very broad range of devices, from powerful PDAs and Java-enabled phones with color screens to rather minimalistic devices with no displays. One challenge for m-commerce is to develop applications and services that can gracefully adapt to the features supported by such a heterogeneous set of devices. Today, many wired Internet Web site developers already struggle to create content that can be properly displayed on both Netscape Navigator and Internet Explorer. Given the variety of mobile platforms, standards are even more critical on the mobile Internet. This is discussed in particular in Chapter 4, where we introduce WAP, a standard aimed at isolating developers from the idiosyncrasies of different mobile devices and communication systems. Java and the 3rd Generation Partnership Project's (3GPP) Mobile Execution Environment (MExE), also discussed in Chapter 4, are motivated by similar concerns. A number of related interoperability challenges are discussed throughout the book, including, for instance, solutions aimed at bridging differences between different location-tracking technologies (Chapter 7).

Convergence of Mobile Telecommunication Networks and the Internet

The explosion in mobile phone ownership in the 1990s was fueled by telephony applications and the appeal of being able to place phone calls "from anywhere at anytime." While voice will remain the main engine of growth in a number of countries, it will soon be overtaken by data traffic in the more mature markets in Europe, the Far East, and North America. Ericsson predicts that, by 2004, there will be over 600 million mobile Internet users. The UMTS Forum, a consortium of 250 organizations that includes mobile operators, vendors, content and service providers, expects that around the same time, data traffic revenues will start dominating mobile voice revenues (see Figure 1.8).

As one would expect, bringing the Internet to mobile phones and wireless PDAs is far from straightforward. This is due to the many limitations of these devices, which include:

- Little memory
- Low transmission speeds
- Frequent disconnects
- Small displays
- Tiny keyboards, if any
- The many different wireless communication standards over which data has to be transmitted

A number of proprietary systems and open standards have been proposed to accommodate these constraints. Early solutions have included Palm's proprietary Web Clipping system, which is designed to selectively download parts of Web pages, and the Handheld Device Markup Language (HDML) developed by Unwired Planet, which was later renamed Phone.com before being eventually acquired by OpenWave. HDML first introduced the "deck of cards" metaphor, where multiple small screens ("cards"), each corresponding to a single user interaction such as selecting from a menu or typing a password, are transferred at once to avoid long waits as the user moves from one screen to the next. HDML eventually evolved into the Wireless Markup Language (WML), the original presentation language of the Wireless Application Protocol (WAP).

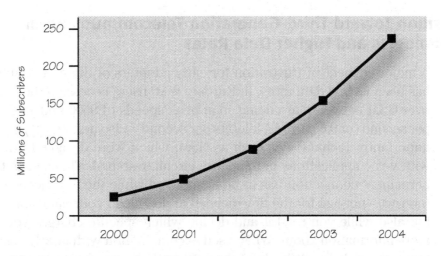

Figure 1.8   Forecast of number of mobile Internet users.

Source: the Allied Business Intelligence/Reuters Insight.

While it has been the subject of much criticism, WAP is quickly evolving into a de facto standard—we discuss this in Chapter 4. Another popular standard is Compact HTML, which was adopted by DoCoMo for its i-Mode service. As we will see, while WAP and i-Mode have been pitted as competitors in the press, they are eventually expected to converge, with both standards already moving toward the adoption of Compact XHTML and better integration with the TCP/IP suite of protocols. Languages such as Java will also play an important role with the emergence of more powerful devices.

Beyond new standards, the mobile Internet is also about a very different user experience. It is characterized by goal-oriented activities, such as reserving movie tickets or looking for directions. These activities are often conducted when under time pressure, such as knowing that movies start at 9 P.M., and are subject to distractions such as talking to friends. Developing services that accommodate these constraints requires moving away from wired Internet solutions such as search engines that return hundreds of hits and Web pages that are full of flashy ads. Instead, what is needed is a move toward transaction-oriented interfaces capable of providing timely and to-the-point responses and dialog-based modes of interaction with the users to help them refine their queries. In Chapter 7, we look at techniques for personalizing mobile Internet services and tailoring their responses to our position as well as to the broader context within which we are operating.

Transition Toward Third-Generation Telecommunication Technologies and Higher Data Rates

A major element of frustration for early adopters of the mobile Internet has been its low data rates. In Europe, WAP users accessing the Internet over GSM networks are doing so at peak speeds of 9600 and 14,400 bits per second (or 9.6 and 14.4 kilobits per second—also denoted *kbps*). More importantly, technologies such as GSM, which were designed primarily with voice applications in mind, are circuit-switched. This means that a permanent connection has to be maintained while the user accesses the Internet—not just for the time needed to download your bank statement, but also while you read it and decide which specific charges you want more information about. When used in combination with poorly designed services that know little about their users, these second-generation cellular technologies can provide for a rather horrendous and potentially expensive user experience, as reported in many user surveys.

The first few years of this decade are being marked by the introduction of faster packet-switched cellular network technologies. In contrast to circuit-switched networks, which require maintaining a connection for the entire duration of an Internet session, packet-switched technologies chop data into small chunks or packets. These packets are routed according to protocols similar to those of the Internet and do not require maintaining a permanent connection, making them much more efficient when it comes to network capacity utilization. With packet-switching, a user can remain connected to the Internet all the time without holding any network capacity and while only paying for the actual number of packets sent and received by his or her device.

Third generation (3G) generally refers to mobile network technologies capable of supporting at least 144 kbps under a broad range of conditions. Just as there are a number of second-generation technologies today (for example, GSM, cdmaOne, TDMA), several families of 3G technologies are emerging. Some will have peak data rates in excess of 2 million bits per second (2 Mbps). This is essentially two orders of magnitude faster than their second-generation ancestors and should allow for a significantly more pleasant Internet experience along with the introduction of videostreaming applications and services, rich interactive games, and so forth. Issues of backward compatibility with 2G technologies, the costs of different upgrades, and the availability and allocation of additional spectrum required for some 3G technologies will generally determine the migration paths chosen by different mobile operators.

In most cases, transitioning from 2G to 3G will not happen overnight. Figure 1.9 depicts some of the main migration paths adopted by mobile operators across the world. In Europe, many mobile network operators have upgraded or are in the process of upgrading their GSM networks to GPRS, an intermediate 2.5G technology that can support packet-switched data transmission at peak rates in excess of 100 kbps—closer to 40 kbps in practice. In June 2000, BT CellNet was the first operator to upgrade to GPRS, although initially the service was only available to corporate users. The next upgrade for many European operators will be one to EDGE or to UMTS, the official European standard for 3G. EDGE, which, coming from GPRS, only involves the deployment of new modulation technology, will have top data rates of 384 kbps. As such, it is viewed by some operators as a cheaper alternative to UMTS, which involves securing new spectrum and much more substantial infrastructure investments. A number of U.S. operators using the TDMA standard,

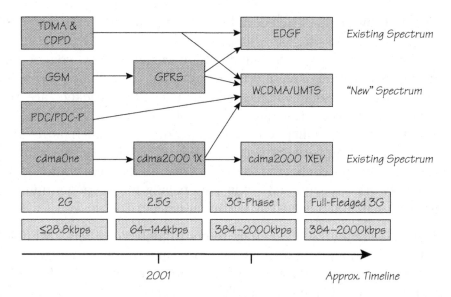

Figure 1.9 Transition toward third-generation standards: A simplified view.

a standard that shares some similarities with GSM, have also decided to upgrade their networks using EDGE technology. In Japan, NTT DoCoMo is upgrading its PDC/PDC-P network to WCDMA—essentially the same 3G standard as Europe's, just with a different name. Finally, other operators in Korea, Japan, and the United States who are already using more efficient CDMA technology are looking at a somewhat easier and less expensive migration to the faster data rates associated with 3G.

If this soup of acronyms sounds intimidating, do not worry. In Chapter 3, "Mobile Communications: The Transition to 3G," we take you through the different standards competing in the marketplace today, and review their main technical characteristics, costs, and adoption levels. As we will see, deployment of 2.5G and 3G is a high-stake game for mobile operators, infrastructure providers, handset manufacturers, service providers, and governments, representing hundreds of billions of dollars in potential combined revenue. While the transition to 3G has been marred by technical delays, unavailability of handsets, and skyrocketing licensing costs, it is clear that the faster data rates it promises are on their way, along with the many new applications and services they will make possible. As can be expected, deployment of 3G will first take place in developed countries. The UMTS Forum predicts that by 2010, only 28 percent of the worldwide population of mobile users, which by then should be around 2.25 billion, will be served over a 3G network, suggesting that intermediate so-called 2.5G technologies will continue to play a key role in the years to come.

Wireless LAN technologies such as IEEE 802.11 and Bluetooth are also expected to play an important role in the uptake of mobile commerce, including, for instance, in the context of mobile point-of-sale (POS) payment solutions, which we discuss in Chapter 5. Finally, while the world is abuzz with 3G, infrastructure equipment providers have already started work on the development of 4G technologies with peak data rates in excess of 100 Mbps, which they hope to start deploying by the end of this decade.

Explosion of Personalized, Location-Sensitive, and Context-Aware Applications and Services

One of the most popular services of the early mobile Internet is the Short Message Service (SMS), which makes it possible for users to send and receive text messages of up to 160 characters. Today, SMS accounts for a substantial percentage of the revenue of some mobile operators. The GSM Association expects that there will be over 200 billion SMS messages exchanged in 2001—there were 15 billion in December 2000 alone. As we have seen in Figures 1.1 and 1.2, mobile operators such as DoCoMo offer a variety of other services to their customers. They range from mobile banking, mobile directory services, mobile ticketing, all the way to mobile entertainment services such as network games, dating services, or downloading of ring tones and screensavers.

While on the surface some of these services might appear similar to those found over the wired Internet, the reality is that mobile devices demand and offer opportunities for much greater personalization than their wired counterparts do. The need for personalization stems from the input/output limitations of the devices and the time-critical nature of the services they are meant to support.

NOTE A student of mine recently reported spending more than 50 minutes purchasing a book from Amazon using her mobile phone. The service did not recognize who she was and asked her to re-enter all her personal details, name, street address, credit card number, and so forth.

Most customers would have given up after a few minutes. Surveys, in fact, suggest that every additional click required from a user reduces by 50 percent the probability of a transaction. Services requiring tedious interactions with their customers will be unlikely to see much business. The key here lies in the ability to tailor services to the user's profile—his or her personal details, preferences, and so forth. Given the reluctance of consumers to enter the same personal information over and over again, this information will be the object of dire competition among m-commerce players. Mobile operators, independent mobile portals, and even banks are all eager to become the primary repository for your personal information, your preferences, location, activities, and billing details. The idea is that they could then convert this intimate knowledge of their customer into favorable revenue-sharing agreements with large numbers of content providers. This business model clearly raises questions about the control the consumer has over the use of his or her personal information, an issue that has not escaped the attention of consumer protection and privacy advocacy groups as well as regulators.

In summary, mobile phones stand to become the most personal computing devices we use. They will increasingly know about our personal preferences and a number of other intimate details such as our whereabouts, activities, friends, and banking information. Armed with this knowledge, they will give us access to context-aware services adapted not just to our location but to a number of other contextual attributes such as the activities we are engaged in, the people we are with, or the weather outside. Chapter 7 provides an in-depth discussion of the technologies, services, business models, and regulatory issues relating to personalization and the provision of location-sensitive and context-aware services.

What's So Special about M-Commerce?

The drivers we just reviewed will likely contribute to making m-commerce what could be a market worth tens of billions of dollars by the end of the decade—imagine over a billion mobile customers each spending an average of $10 per month. However, isn't all this just good old e-commerce? Certainly, m-commerce does involve the buying and selling of goods and services over the Internet. The answer is that, just like e-commerce, m-commerce is more than just another distribution channel with its own idiosyncratic technologies. M-commerce is also about the introduction of unique services, usage scenarios, business models, and regulatory challenges. Ignoring these differences is a recipe for failure. In the next section, we summarize some of the key differentiating factors of m-commerce.

The Usability Challenge

Surveys of mobile Internet users indicate that usability is by far the biggest source of frustration among early adopters. Their complaints stem from the many limitations of today's mobile devices: limited screen *real estate*, limited input functionality in the form of tiny keyboards, and limited memory and processing power. They also reflect frustration with the low data rates and frequent disconnects associated with today's mobile Internet experience. In contrast to the flashy and colorful sites found on the wired Internet, developing successful m-commerce services requires a much more disciplined approach, where well-targeted content is presented to the user in as concise a form as possible. Forget about the sophisticated animations and banner ads of e-commerce. Forget also about search engines that return hundreds of hits and Web sites with five-page privacy policy statements.

Addressing the m-commerce usability challenge can only be done by rethinking all aspects of the user interaction. This involves moving away from a paradigm where more is better and where masses of information are thrown at the user in the hope that some of it *sticks* and catches his or her attention. Personalization is the name of the game and, while cookies, customer relationship management, and collaborative filtering all have their roots in wired e-commerce, personalization takes a whole new dimension in the mobile Internet. It involves presenting customers with services that are relevant to their current locations, activities, and surrounding environments. This means telling you about the ice cream parlor at the corner when it is sunny and you are strolling down the street with your kids, or telling you about the Kinko's outlet at the other end of the street when you need to print that all-important document you have to hand to your client in just half an hour. It also means moving away from dumb menus and developing smarter interfaces capable of entertaining dialogs with users, learning from their behaviors and leveraging prior knowledge of their many preferences. Finally, it involves using other modes of interaction such as speech-based interfaces capable of interpreting voice commands.

Usability is one of the key themes of this book—there is not one chapter that does not touch on it. We discuss it in the context of technologies such as WAP or VoiceXML. We look at its relationship with different business models. We discuss it in Chapter 7, where we review personalization and context-awareness issues.

New Usage Scenarios

The mobile Internet opens the door to a number of services and applications that would simply be inconceivable from a desktop PC. They include new ways of staying in touch with others while on the move, such as sending SMS messages, checking for nearby people with similar interests such as potential partners for a mobile multiplayer game, or accessing the company's intranet in order to complete the sale of a new insurance policy. Banking on your mobile phone, looking for the nearest pizza restaurant, or buying a coke using your mobile phone are yet other examples of the many new usage scenarios associated with the mobile Internet. Many of these revolve around time-critical needs that require short, to-the-point interactions. The time criticality of the task outweighs the limitations of the access device. You probably would not want to use your mobile phone to engage in a complex Internet session aimed at fine-tuning your retirement portfolio, but you would much more readily use it to check your favorite stock or find out about the departure time of the next flight home (see Figure 1.10).

M-commerce is about identifying these time-critical tasks and developing compelling services and profitable business models to address them. Many experts believe that we have only started to scratch the surface of m-commerce, and that many more services and usage scenarios remain to be invented. The more futuristically inclined among them envision pervasive computing scenarios where new technologies for personalization and context-awareness and new modes of interaction will make it possible to identify our every wish and present us with adaptive services aimed at facilitating all aspects of our daily lives. Emerging usage scenarios are another key theme of this book. Chapter 8, "Early Lessons and Future Prospects," also includes a brief discussion of where m-commerce might be headed five to 10 years down the road, based on insight from research projects that leading companies, and universities across the world are engaged in today.

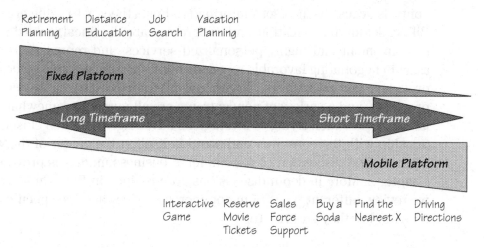

Figure 1.10 Mobile Internet usage scenarios: when time criticality outweighs the device's limitations.

New Business Models

M-commerce is changing the role of most players across the telecom value chain, while also providing room for many new entrants. In the process, new business models are being invented. While each player is experimenting with its own variation, one thing is clear: mobile commerce often involves a complex web of business partners, from technology platform vendors to infrastructure equipment vendors and handset manufacturers, all the way to application developers, content and service providers, mobile telecom operators, banks, content aggregators, and mobile portals, to name just a few. The result is an m-commerce jigsaw puzzle where each player can only hope to focus on a small subset of the entire value creation process.

Mobile operators view m-commerce as an opportunity to reinvent themselves as sophisticated high-margin transaction support providers, service providers, and content aggregators. The success of this transformation depends to a great extent on their ability to leverage their existing customer relationship and build partnerships with a critical mass of content providers. At the core of this strategy is the realization that whoever controls access to the user's information holds the key to value creation. Billing details, user location, and preferences are all critical to the delivery of convenient and highly personalized services, and represent a major asset in negotiating favorable revenue-sharing arrangements with content providers and advertisers. Traditional Internet portals, banks, content providers, and a variety of new entrants are all pursuing somewhat similar strategies, each trying to leverage its advantage to build a closer relationship with the consumer and position itself at the center of the value chain. A first discussion of m-commerce business models is provided in Chapter 2. More in-depth discussions are provided in Part Three, where we review different types of m-commerce services and the specific business models they give rise to.

Interoperability Challenge

M-commerce marks the convergence between a culture of openness and interoperability, which has been the Internet's hallmark, and a world fragmented into pockets of competing mobile telecommunication standards. Economies of scale, complex value chains, and consumer demand for global roaming and affordable services all require interoperability. Reconciling the many standards and technologies of the emerging m-commerce industry, from data transport to billing, location tracking, and payment, is no easy task. Throughout this book, we discuss a variety of efforts that aim at doing just that.

Security and Privacy Challenges

Security and privacy are yet another area where m-commerce imposes new constraints. Vulnerability of the air interface, the limited computing power of mobile devices, and the low data rates and frequent disconnects of wireless communication all provide for challenging problems when it comes to guaranteeing end-to-end security. As mobile devices proliferate and start being used to access corporate intranets and extranets, they also become increasingly tempting targets for hackers. However, probably one of the most challenging aspects of m-commerce has to do with finding solutions that reconcile users' demand for highly personalized services with their desire for privacy. There is a fine line between presenting consumers with useful services that are relevant to their location and preferences, and bombarding them with annoying location-sensitive ads. Organizations as diverse as the World Wide Web Consortium (W3C) and the Mobile Marketing Association (MMA) are attempting to define standards that will help preserve the privacy of wireless users and prevent spamming.

Security issues relating to mobile communication standards, Internet standards and mobile payment solutions are respectively addressed in Chapters 3, 4, and 5. Privacy issues are discussed in several chapters, including Chapter 7.

How This Book Is Organized

We have tried to organize the chapters in this book in a modular fashion to accommodate the interests and backgrounds of as many readers as possible. We have also tried to keep a fairly informal style, while doing our best to introduce every single technical term and acronym (by now, if you have read through this first chapter, you should realize that there are quite a few). There is also a glossary of terms provided at the end of the book to save you from having to remember where each term was first introduced—especially if you do not plan to read every chapter. In general, we have assumed a fairly minimal technological understanding. If you are a manager who reads technology articles in publications such as *The Wall Street Journal* or *BusinessWeek*, you should have no problem following and will hopefully enjoy the many business discussions we have included. On the other hand, if you are a developer who is already familiar with mobile communication technologies, Internet protocols, or encryption, you might want to just skim through the first few pages of chapters where we briefly review some of the basics associated with these technologies. In general, we have tried to stay away from deep technological discussions—there are a number of excellent textbooks on each of the technologies we cover, some of which are listed in the references at the end of this text. Instead, our goal is to give you a broad overview of the many technologies, services, and business models of mobile commerce. In the process, we try to show how technologies and standards impact the business models of players across the value chain, and how existing and new business models and usage scenarios also drive the development of new technologies and standards.

Specifically, the book is organized in three parts.

Part I: The M-Commerce Revolution

Chapter 1: M-Commerce: What's the Buzz All About?

Here, we attempt to understand what mobile commerce (m-commerce) is all about. We take a first look at m-commerce services and the business models of different players across the m-commerce value chain. In the process, we also review the drivers behind the emergence of m-commerce, and explain how it differs from its older cousin, e-commerce. Hopefully, as you reach the end of this chapter, you will have realized that m-commerce is actually quite different—not just in terms of technologies, but also in terms of services, usage scenarios, players, and business models.

Chapter 2: A First Look at the Broader M-Commerce "Value Chain"

In Chapter 2, we take a closer look at the many different categories of players found across the m-commerce value chain. We try to understand the context within which they operate, and how the opportunities and threats of m-commerce are already starting to impact their business models and their place in the value chain. This includes a discussion of mobile operators, mobile portals, content providers, Wireless Application Service Providers (WASPs), payment providers, location brokers, advertisers, handset manufacturers, equipment providers, and more.

Part II: The Technologies of M-Commerce

Chapter 3: Mobile Communications: The Transition to 3G

Chapter 3 starts with a brief introduction of basic mobile communication principles, and follows with an overview of the main 2G, 2.5G, and 3G communication standards. Rather than getting into the nitty-gritty of every single standard, our discussion focuses on major features introduced by different generations of technologies, such as packet-switched communication, always-on functionality, and faster data rates. We examine key factors influencing the selection of one standard over another, as different mobile operators upgrade their networks. We also look at the timeframe associated with the transition to 3G and its impact on the development of successful m-commerce applications and services over the years to come.

Chapter 4: The Mobile Internet

Traditional Internet protocols and Web standards developed by the Internet Engineering Task Force (IETF) and the World Wide Web Consortium were not originally designed with mobile communication networks in mind. The mobile Internet is about adapting these standards to the specificity of mobile environments and the multitude of 2G, 2.5G, and 3G mobile communication technologies found around the world. This chapter attempts to introduce some of the basic issues involved in reconciling two fairly different views of the world: the Internet view, and the mobile communication view. As we will see, much progress has been made in this area over the past few years, with Internet protocols being adapted to the demands of mobility and with the emergence of a de facto standard, the Wireless Application Protocol (WAP), which has recently made significant efforts to converge with IETF and W3C standards. The situation, however, is far from perfect, and security issues, for one, continue to require special attention. These and other considerations have important implications for players across the entire m-commerce value chain.

Chapter 5: Mobile Security and Payment

While issues of security are addressed in Chapters 3 and 4, this chapter revisits some of them. In particular, we provide an informal review of basic cryptography mechanisms used to support security—no, there will not be a single mathematical equation in the text—and examine how these mechanisms are affected by the limitations of mobile devices and mobile communication networks. This includes a discussion of both secret and public key cryptography, and the increasingly important role of smart cards to support security on mobile devices. The second part of this chapter is devoted to mobile payments, which rely heavily on the mobile security solutions discussed earlier in the chapter. As we will see, mobile operators are trying to leverage their control of authentication mechanisms and their micropayment infrastructures, to become the mobile user's preferred payment provider. We review solutions and emerging standards that aim at circumventing this situation, as well as mobile payment solutions that extend approaches originally developed for the fixed Internet. Some of these efforts include using mobile devices as part of point-of-sale payment solutions, where they could replace cash and credit cards

Part III: M-Commerce Services Today and Tomorrow

Chapter 6: Mobile Commerce Services Today

Part Three is where all the pieces come together. We begin by looking at m-commerce services available today. Our discussion covers both consumer services and business applications. Here, we review mobile portals as well a number of more specific services such as mobile banking, mobile ticketing, or mobile entertainment. In each case, we look at emerging usage scenarios and business opportunities and compare the approaches taken by different players. The chapter ends with a discussion of solutions aimed at empowering mobile workers by enabling them to remotely connect to the enterprise's intranet and access a variety of business applications such as mobile sales force support, mobile ERP, mobile CRM, and mobile fleet management.

Chapter 7: Next-Generation M-Commerce: Context Awareness and Interoperability

Many of today's m-commerce services remain very limited when it comes to offering customers services that are directly relevant to their preferences, location, or other contextual attributes (for example, weather, people you are with, activities you are trying to accomplish, and so forth). In this chapter, we introduce new solutions aimed at facilitating the personalization and contextualization of mobile services. This includes discussion of 3GPP's Personal Service Environment, which aims at providing a single repository for all the customer's personal preferences, as well as similar efforts recently launched by industry leaders such as Microsoft or members of Sun's Liberty Alliance. This is followed by a presentation of new positioning technologies and the many location-sensitive services they entail. Here again, we look at the different players in the market and their business models. We conclude with a discussion of future context-aware solutions, based on ongoing research efforts in industry and academia.

Chapter 8: Early Lessons and Future Prospects

3G, WAP, and the mobile Internet have all been the subject of a lot of hype over the past few years. This is hardly a surprise, given the hundreds of billions of dollars at stake in infrastructure equipment, spectrum licenses, and projected revenue opportunities—not to mention the

Internet gold rush mentality that characterized the late 1990s. Since then, technological glitches, user complaints, poorly thought-out business models, and a number of other setbacks have forced many players to re-evaluate their forecast and strategies. In this chapter, we review some of the early lessons of mobile commerce and attempt to provide more realistic predictions for the years to come, outlining both short-term and long-term opportunities and challenges.

Further Readings

For those of you interested in further exploring some of the topics discussed in this text, we have included a list of references and Web pointers at the end of the book.

CHAPTER 2

A First Look at the Broader M-Commerce Value Chain

Introduction

Before we delve into the many technologies, applications, and services of m-commerce, it helps to have an overall view of the m-commerce value chain. As we saw in Chapter 1, "M-Commerce: What's the Buzz All About?", m-commerce is generally characterized by a variety of business partnerships often involving a large number of organizations, from content providers to WASPs to mobile network operators. While no two value chains are the same, the objective of this chapter is to introduce you to some of the main categories of players involved in the creation and delivery of m-commerce applications and services, to help you understand the role they each play and the context within which they operate. This chapter also provides a first discussion of major m-commerce sources of revenue and business models. In Part Three, we will refine these generic models and look at some of their more specific instantiations in the context of different types of applications and services.

As Figure 2.1 illustrates, the delivery of m-commerce applications and services involves a number of players. They generally fall into one or more of the following categories:

- Infrastructure equipment vendors
- Software vendors

- Content providers, including advertisers
- Content aggregators
- Mobile network operators (MNOs), including virtual ones (VMNOs)
- Mobile portals
- Third-party billing providers
- Mobile device manufacturers
- Wireless Application Service Providers (WASPs)
- Location information brokers

Infrastructure Equipment Vendors

While they are not very visible to the consumer, infrastructure equipment vendors play a key role in m-commerce. They provide the base stations, mobile switching systems, and other solutions necessary for the wireless transmission of voice and data. In 2000, this market was already worth

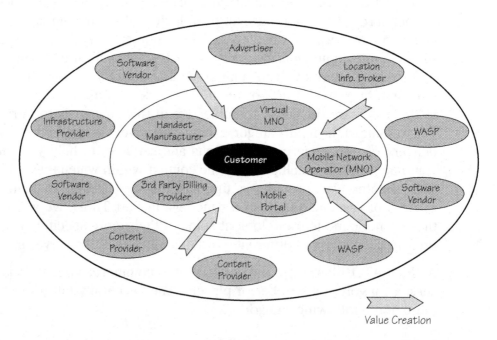

Figure 2.1 The m-commerce value chain involves a complex web of actors only some of which are shown in this generic illustration.

around $50 billion. According to research firm Forward Concepts, as the number of mobile phone users continues to grow and as operators world-wide start deploying 2.5G and 3G technologies, the mobile infrastructure equipment market could very well double by 2005. In fact, while prices paid by a number of mobile operators for 3G licenses made the headlines, the cost of deploying the required infrastructure might prove to almost be on the par with that paid for the licenses, possibly reaching as much as $100 billion over the years to come in Europe alone. With such huge sums at stake, it should be no surprise that equipment vendors have been among the most ardent proponents of the mobile Internet and m-commerce. With a number of mobile operators strapped for cash following the purchase of spectrum licenses, they also found that they had no choice but to consent to major loans and other financial incentives to entice operators to go ahead with 2.5G and 3G deployment plans. In the process, they have even more closely tied their fate to that of mobile operators, and find themselves to be major stakeholders in the m-commerce market. The emergence of just three families of 3G standards—WCDMA/UMTS, EDGE, and cdma2000 (see Figure 1.9)—is also breaking down technological barriers between regional markets and intensifying competition among infrastructure equipment makers.

Key mobile infrastructure vendors include Ericsson, which in 1999 held close to 30 percent of the worldwide market, followed by Lucent, Motorola, which is working closely with Cisco, and Nokia, each with market shares around 10 percent at the time. Other major players include Nortel, NEC, Siemens, Alcatel, Fujitsu, Samsung, and Qualcomm, whose cdma technology forms the basis of the two main 3G standards, WCDMA/UMTS and cdma 2000. As these companies battle for shares of the huge infrastructure market, they each benefit from different advantages. Significant barriers to entry, such as major R&D investments and fears of interoperability problems by mobile operators, generally limit the GPRS infrastructure market to major GSM infrastructure players such as Ericsson and Nokia. With a combined control of over 50 percent of the GPRS infrastructure market so far, both companies are well positioned to secure a substantial share of EDGE and UMTS/WCDMA upgrades. Ericsson and Nokia have also secured a number of early WCDMA contracts in Asia, including Ericsson's key contract to supply WCDMA infrastructure equipment to NTT DoCoMo as it launches FOMA, the first 3G service in the world. Similarly, well-established vendors of CDMA technologies such as Nortel, Motorola, Lucent, Samsung, and Ericsson, through its cross-licensing agreement with Qualcomm, stand to fare well when it comes to upgrading cdma networks.

As major mobile Internet stakeholders, infrastructure vendors are also playing a critical role in key standardization and interoperability initiatives, including:

- The **Third Generation Partnership Projects**, 3GPP and 3GPP2, which aim at developing global specifications for third-generation mobile systems based on evolutions of GSM and CDMA technologies, respectively.
- The **Wireless Application Protocol** (WAP) Forum, which works on the development of mobile Internet standards.
- The **Mobile electronic Transaction** (MeT) initiative, which focuses on the development of standards for secure mobile transactions.
- The **Location Interoperability Forum** (LIF), whose objective is to provide for interoperability across different location-tracking technologies.

Finally, another emerging segment of the wireless infrastructure market is the one of wireless local area network (WLAN) technologies such as IEEE802.11 and Bluetooth, with infrastructure sales of over $1 billion in 2000 and growing fast. Over time, these technologies will likely be integrated into the 3G infrastructure, making it possible for someone to seamlessly access m-commerce services and applications across both wide area (cellular) networks and WLANs; for example, as users enter malls or airports, their devices will automatically connect through the WLAN.

Software Vendors

Software vendors are another important, yet not very visible, part of the broader m-commerce value chain. They are the suppliers of operating systems, databases, microbrowsers, and other middleware technologies that are central to providing a secure and user-friendly experience to the mobile customer. As mobile operators, handset manufacturers, mobile portals, and other m-commerce players pick among different software platforms, they inevitably limit the number of partners they can work with and services they can offer, due to numerous incompatibility problems across competing offerings.

In the operating system arena, major contenders include:

- EPOC, developed by the Symbian consortium, which brings together companies such as Psion, Motorola, Ericsson, Nokia, and Matsushita.
- Windows CE and its Singer version specifically developed by Microsoft for mobile phones.
- PalmOS, which currently runs on over 60 percent of all PDAs.

The microbrowser war is dominated by Openwave (formerly known as Phone.com), Nokia, Microsoft, and Ericsson, and also includes new entrants such as 4thPass with its Java-enabled microbrowser. Key players in the mobile database market include Sybase subsidiary, iAnywhere Solutions, Oracle, IBM and a small number of other contenders. Because of the variety of software solutions and the number of players in each market, a complete overview of this category of players is beyond the scope of this book. As can be expected, most of these players rely on business models that combine licensing, consulting and maintenance fees. A growing segment of these companies also operate as WASPs. This is further discussed in the Section *Wireless Application Service Providers (WASPs)*.

Content Providers

Above all, m-commerce is about content and giving users access to a myriad of mobile services. As we already saw in Chapter 1, mobile services fall under a number of categories, giving rise to different possible sources of revenue and business models. Content can range from news to directory services, directions, shopping and ticketing services, entertainment services, financial services, and so forth. Possible sources of revenue include subscription fees, transaction fees, share of traffic charges collected by the mobile operator, and various forms of sponsorship such as advertising, referral fees, and commissions. They can be combined in a number of different ways. However, one can generally distinguish between the following core business models:

- User Fee Business Models
    - Subscription fees
    - Usage fees

- Shopping Business Models
- Marketing Business Models
- Improved Efficiency Business Models
- Advertising Business Models
- Revenue-Sharing Business Models

User Fee Business Models

What better way to make money than to charge users for the content they access? This model works as long as the content is compelling enough and it is regularly updated. Typical content that users will be willing to pay for can range from news, to traffic conditions, games and entertainment, to highly specialized information such as weather conditions at different golf clubs around town. Payment can be in the form of a subscription fee, on an actual-usage basis, or even a combination of both.

Subscription fees tend to be easier to collect than transaction fees, and provide a more predictable source of income. Independent content providers might find it more economical to rely on the mobile network operator for the collection of subscriptions, as in the current i-Mode model where DoCoMo keeps 9 percent of the fee it collects on behalf of official content providers (see Figure 2.2). A typical example of a content provider relying on this model is Bandai and its Chara-Pa service, which, for a dollar a month, allows customers to download one cartoon a day as a screensaver (see Figure 2.3). By April 2000, the

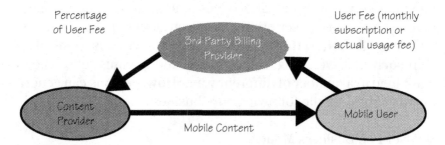

Figure 2.2 The User Fee Business Model often involves relying on a third-party micro-billing provider, whether for the collection of subscription fees or for actual usage fees.

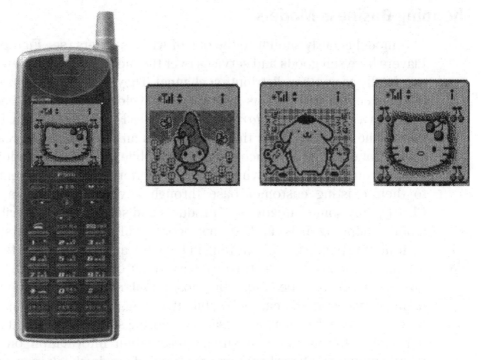

Figure 2.3 Bandai's Chara-Pa service allows subscribers to download one screensaver a day as a GIF file for a monthly fee of $1.

service already had 1.6 million subscribers. As of November 2000, 30 percent of all official i-Mode content providers were charging a monthly subscription fee.

Usage fees are the ultimate source of revenue in the sense that they make it possible to charge for actual usage of a service, with each access generating additional revenue for the content provider. With the majority of mobile transactions involving small fees (for example, 50 cents for driving directions or for a traffic update, or a dollar to download a song), this model is even more difficult for small content providers to implement on their own than the subscription model, requiring again that they partner with a mobile operator, a mobile portal, or some other third-party micro-billing provider. This is discussed further in the section *Third-Party Billing and Payment Providers*.

Most of the sites that charge user fees also provide some content for free to entice customers to subscribe to their premium content.

Shopping Business Models

This model is fairly similar to the one of wired e-tailers (see Figure 2.4). Players here sell goods and services over the mobile Internet, viewing it essentially as another distribution channel. They include both pure-play Internet companies such as Amazon or Travelocity, as well as companies with a *brick-and-mortar* presence such as Fleurop Interflora. Using the mobile Internet offers these companies an opportunity to reach a somewhat different audience—so far the mobile Internet population has proved younger than the wired one, while offering added convenience to their existing customer base through anywhere/anytime access. Clearly, only some categories of products and services are amenable to mobile shopping. It is unlikely that people will start buying cars from their mobile phones, while waiting in the subway. Buying movie tickets, arranging for last-minute travel reservations or purchasing CDs and flowers are likely to be among the most popular categories of mobile e-tailing services. In the case of mobile ticketing, EMPS, a joint initiative between Nokia, Nordea, and Visa, is exploring solutions that will make it possible for users to download reservation information on their mobile phone and later beam it at airport and movie check-in counters, using technology such as Bluetooth. As more autonomous forms of mobile shopping become available through the launch of shopping agents (or *shopbots*), we might see a broader range of items purchased over the mobile Internet. Sites such as Yahoo!Mobile already offer wireless access to comparison shopping engines that search and compare prices among merchants registered with the portal.

Figure 2.4 The Shopping Business Model is similar to the one found on the wired Internet. Payment also often involves a third party, not represented here, such as a credit card company, bank, or mobile network operator, which will generally keep a percentage of the transaction.

However, the mobile Internet shopping model is not just about over-coming the limitations of mobile devices. It also opens the door to much more convenient and more personalized, location-sensitive, and con-text-aware scenarios, where mobile users are presented with products directly relevant to their current locations or activities. This includes telling mobile users about nearby hotels if they are on a business trip, or nearby movies if they are looking for something to do in the evening. In general, because they are so personal and follow you wherever you go, mobile devices are an ideal marketing channel for impulse buying. This is discussed further in the sections *Marketing Business Models* and *Advertising Business Models*. The mobile Internet also offers the possi-bility of anytime, anywhere extension to wired online auctions, making it possible for users to enter and follow auctions while on the move. Today, major online auction houses such as eBay already support wire-less access to their sites. Finally, mobile shopping combined with loca-tion-tracking functionality gives vendors the ability to automatically determine where a service or product should be delivered. This could prove particularly appealing to taxi companies or pizza delivery outlets.

Note also that, while mobile shopping transactions tend to involve larger sums of money than those in the User Fee Business Model, they also gen-erally involve third-party billing through credit card companies, banks, or mobile operators.

Marketing Business Models

Brick-and-mortar companies and traditional online players can decide to simply use the mobile Internet as a marketing channel for their core business. Under this model, the company creates a mobile presence to reach existing and potential customers, but does not necessarily sell anything over the mobile Internet. In other words, the mobile presence is subsidized by the company's core business, which can range from sell-ing cars, CDs, or magazines, to renting videos or even offering education programs in the case of a university (see Figure 2.5).

Early experience with mobile marketing, as reported by companies such as Tsutaya, a video rental and CD retail chain with 1,000 outlets across Japan, suggests that, if properly used, it can be a very effective tool. For example, Tsutaya lets its users enter their music preferences and sends them notifications when their favorite artists release new CDs or are scheduled to give concerts. As of late 2000, Tsutaya's mobile marketing

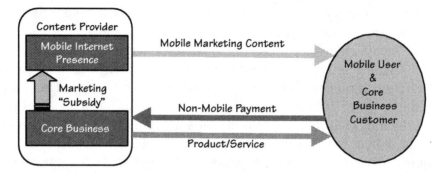

Figure 2.5 The Marketing Business Model is adopted by brick-and-mortar players and traditional Internet companies interested in using the mobile Internet as a marketing channel rather than as an actual sales channel. Their core business subsidizes their mobile Internet presence.

channel was reporting over 650,000 users and over 2 million accesses per week.

The key to effective mobile marketing is knowing your customer and leveraging that knowledge to deliver highly relevant messages; namely, messages that reflect their personal preferences as well as possibly their locations or other contextual attributes. Given the limited screen real estate available on mobile devices and the limited attention span of users, failure to deliver well-targeted messages can potentially backlash and antagonize users. Privacy laws and emerging anti-spamming legislation further protect consumers from unwanted marketing messages. Permission marketing approaches such as Tsutaya's are often an effective way to obtain information about the user along with their consent to receive a certain number of notification messages each day. Variations of this model include offering users to enroll in lotteries, giving away trial subscriptions to services, and so forth. Often, initial contact with users takes place over the wired Internet, where they are prompted to provide their personal details. This information is later used to send them targeted promotional messages on their wireless devices. This includes the delivery of online coupons that can be saved on the device and later redeemed at retail outlets. Tsutaya has reported that, among its customers, mobile users who had received online coupons were 70 percent more likely to visit their stores and, on average, spent 59 percent more than their *non-mobile* counterparts. We report similar results later in the chapter, where we discuss Mobile Advertising Business Models.

Improved Efficiency Models

Other content providers simply view the mobile Internet as an opportunity to cut costs and improve customer satisfaction. This is similar to the view many companies have of the wired Internet, where an online presence can help reduce operating expenses. Mobile examples of this model include mobile banking, mobile trading, or mobile ticketing. These solutions make it possible for companies to cut down on personnel at branch offices, call centers, ticketing booths, and counters. They can also help disintermediate traditional business models, as in the case of airlines selling tickets directly to consumers instead of going through travel agencies. All in all, conducting business over the mobile Internet can result in significant cost savings, offsetting the cost of *mobilizing* business processes. Nordea's WAP Solo banking solution, which we introduced in Chapter 1, falls into this category. It helps make the company's operation less human intensive, while offering customers the added convenience of anywhere, anytime access to a number of banking, trading, and shopping services (see Figure 2.6).

In Japan, Daiwa Securities reported in late 2000 that 35 percent of its customers bought and sold stock over the Internet, with 20 percent of these transactions taking place over i-Mode, accounting for close to 7 percent of all their transactions. Daiwa has also reported that its mobile online transactions were about 50 percent cheaper than traditional ones.

Similar cost-saving and efficiency arguments can be made for business-to-business scenarios and business-to-employee scenarios, where *mobilizing*

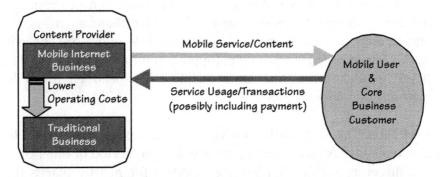

Figure 2.6 The Improved Efficiency Business Model: savings in operating costs and added convenience to the user offset the costs of setting up a mobile presence.

the workforce and enterprise systems can help boost company productivity. Examples range from mobile groupware solutions to mobile sales force support applications that allow salespeople to access enterprise back-end systems, check product availability, and negotiate more competitive delivery dates with customers.

Advertising Business Models

Advertising, the model of choice in the early days of the wired Internet, has shown its limitations, at least when used as the sole source of revenue. Yet it remains a valid model, offering content providers a valuable source of extra revenue. As already mentioned in our discussion of the Marketing Business Model, the small screen size of mobile devices requires a much more targeted approach to advertising than on the wired Internet. This can be done by presenting users with ads that are directly relevant to queries they enter—when the user is looking for a place to eat, present him or her with coupons for nearby restaurants. It can also involve collecting users' preferences or developing solutions that will deliver ads that are sensitive to their current locations and other contextual attributes such as time of day or local weather conditions. If it is sunny, tell the user about the ice cream place around the corner. Finally, different types of incentives can help make users more receptive to promotional messages. This can include discounts on their mobile Internet plans. Some sites such Keitai Net in Japan go as far as paying users 15 yens per ad they click on.

Figure 2.7 presents a generic illustration of the Advertising Business Model. The advertiser generally pays a fee to the content provider for adding promotional messages to the content it delivers to mobile users. In practice, a number of variations of this model can be found, many involving intermediaries such as wireless advertising agencies, content aggregators, mobile portals, mobile network operators, or wireless ASPs. Advertising fees can be computed in different ways. While many hybrid models exist, one often distinguishes between the following methods of computing fees:

Flat fees. The simplest form involves charging the advertiser a flat fee in exchange for showing the ad over a given period of time. Go2.com, a directory service supporting searches for nearby places through specialized queries such as its *go2movies*, *go2restaurants*, *go2doctors*, or *go2atm* services, offers a variation of this model to prospec-

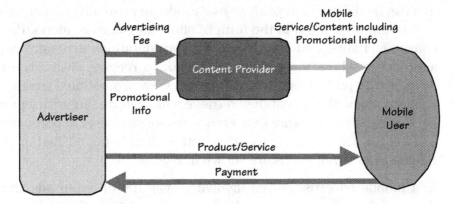

Figure 2.7 Generic Advertising Business Model. Variations might include wireless advertising agencies, working as intermediaries between advertisers and content providers.

tive advertisers. Specifically, it charges them a placement fee of around $20 per month to figure among the companies it will list. Go2.com has also entered larger deals such as the one it has with Coca Cola, where official Coke outlets receive premium placement in its directory service along with driving directions for customers.

Traffic-based fees. Like on the wired Web, traffic-based fees enable advertisers to pay based on the number of times their message is displayed. CPM or *cost-per-thousand* ad impressions is the standard metric being used here. Early wireless ad campaigns suggest that advertisers may be willing to pay substantially more for 1,000 ad impressions on the mobile Internet than on the wired one. This is in part explained by the small screens of mobile devices, as well as the higher success rates reported so far for wireless ad campaigns in comparison with traditional online ones.

Performance-based fees. Just like in the wired Internet space, advertisers will often demand fees that reflect actual results. Fees based on the number of *click-throughs* or *call-throughs*, or more generally, fees that depend on the specific action taken by the user (*cost-per-action*) such as subscribing to a service, as well as commission fees based on actual sales (*cost-per-sale*), all fall under this category. An example of this model is implemented by Japanese content provider GolfOnline. The company receives a commission from partnering golf courses on every reservation made via its site.

Ads can be delivered in *push* or *pull* mode. In push advertising, ads are sent to the user, often in the form of short messages or alerts. Privacy, consumer rights, and user tolerance considerations generally require approaches where the consumer *opts in* to receive push advertising messages. In pull advertising, consumers view promotional messages in combination with content they retrieve while interacting with a particular Web site. Here, ads are delivered in response to user actions, such as looking for nearby movie theatres, rather than being pushed on them. Examples of wireless ads are shown in Figure 2.8.

In a mobile advertising trial involving over 100 different ads and the delivery of 2 million wireless ad impressions in late 2000, wireless advertising company Windwire reported click-through and call-through rates ranging between 10 and 15 percent, in contrast to regular wired Internet click-through rates of 0.5 percent or less. Without a doubt, such high figures partly reflect people's initial curiosity about the ads. At the same time, they suggest that mobile devices are at least as good an advertising medium as the wired Internet and quite possibly a better one. This is in part due to the limited screen sizes, which actually increase the chance of the user seeing any given ad. In the longer run, the ability to send targeted ads based on the user's location or other contextual attributes should further increase the effectiveness of mobile advertising.

In late 2000, research firm Ovum predicted that mobile advertising could grow into a $4 billion industry by 2003, reaching $16 billion by 2005, assuming a population of 500 million mobile Internet users by then. While other research firms are more cautious in their forecast, most

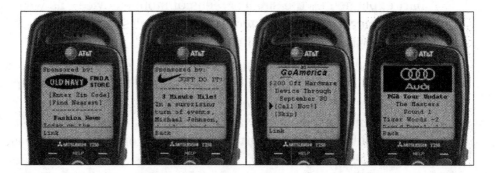

Figure 2.8 Sample wireless ads.

Courtesy of Windwire.

predict that mobile advertising will turn into a lucrative business in the years to come.

Revenue-Sharing Business Models

As we saw in the User Fee Business Model, selling content directly to the consumer is not always an easy proposition, in part due to the marketing overhead and the need to often rely on third-party billing providers. In other situations, the information or services maintained by a company, while valuable, might not be sufficient for a standalone Web presence. Instead, the content provider might have to rely on partnership arrangements with other companies that will combine its content with that of others in order to deliver a compelling service. Examples of content often falling under this category include local weather updates, traffic conditions, news updates, and games and other entertainment services.

Revenue sharing generally involves collecting payment from the user and redistributing it across the different parties involved in delivering the service (see Figure 2.9). Companies such as Webraska, which provides driving maps, along with real-time traffic reporting, complete with alternate route suggestions, rely on revenue-sharing arrangements with mobile operators such as Orange, M1, or Voicestream. Webraska itself has revenue-sharing arrangements with other content providers such as London-based Trafficmaster, which supplies it with up-to-date traffic and travel-time information derived from a private network of bridge-mounted infrared sensors monitoring traffic speed on British roads. Mobile game providers such as Digital Bridges or BlueFactory have similar revenue-

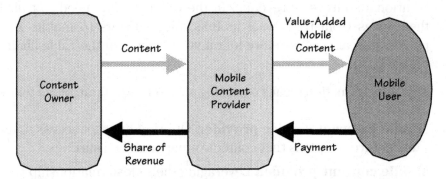

Figure 2.9 Revenue Sharing Business Model. When the mobile content provider happens to be a mobile network operator, payment might be traffic-based.

sharing arrangements with mobile operators. Popular games like Digital Bridges' *Wireless Pet*, where a mobile user has to care for a digital pet, have been reported to already have tens of thousands of users generating close to 100,000 hours of airtime per month for mobile operators across Europe. As we will see later in this chapter, companies relying on revenue-sharing arrangements with mobile network operators often view themselves as wireless ASPs (WASPs). This is the case of companies such as Webraska or BlueFactory.

Content Aggregators

Content aggregators focus on value creation by assembling content from multiple sources. Many mobile content aggregators focus on repackaging information for distribution over the mobile Internet. For example, Singapore-based BuzzCity aggregates, repurposes, and WAP-enables local, regional, and international content through its djuice mobile portal, which is redistributed by mobile operators such as Telenor in Norway, DiGi in Malaysia, and TotalAccess in Thailand. In general, independent mobile portals, most mobile network operators, and many wireless ASPs operate as content aggregators. Each of these categories of players is the subject of a section in the remainder of this chapter.

Mobile Network Operators

With falling profit margins in the mature mobile voice markets of Asia, Europe, and North America, mobile network operators are under increasing pressure to turn to mobile data services and m-commerce for additional sources of revenue. In the process, they need to transform themselves from traditional mobile voice carriers to mobile Internet players that also cover more lucrative segments of the value chain (see Figure 2.10).

This transformation means operating also as one or more of the following:

Mobile Internet service provider. Offering Internet access to mobile users through plans that combine email and Web surfing.

Mobile content provider. Leveraging their close relationship with the customer and their control of the first screen to deliver compelling information and services.

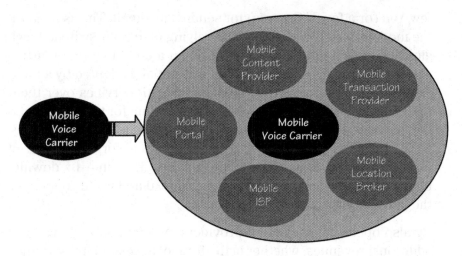

Figure 2.10 The changing role of the mobile network operator.

Mobile portal. Offering mobile users a one-stop solution for all their mobile Internet needs, from personalized content to messaging, calendar, and other Personal Information Management (PIM) applications and, in the process, positioning themselves as primary repository for the user's personal information and preferences.

Mobile location broker. Leveraging information about the mobile user's position to enter into profitable partnership arrangements with content providers, wireless ASPs, and portals.

Mobile transaction provider. Taking advantage of their existing billing relationship with the customer and their micropayment infrastructure to collect fees on behalf of content providers.

The result is a number of possible business models combining different sources of revenue. Most mobile operators now offer Internet access plans to their users for a few dollars a month. In addition to mobile Internet subscriptions, Internet access itself is a source of revenue through traffic and access charges. Companies operating packet-switched networks such as NTT DoCoMo bill customers based on the actual bandwidth they use; namely, the number of packets they send and receive—between 0.2 and 0.02 Yen per 128-byte packet at the launch of DoCoMo's FOMA 3G service, which typically results in charges of just a few Yen (or a few U.S. cents) for sending an email. This is in contrast to the models used by operators still relying on circuit-switched technologies, who often bill their users based on a combination of airtime and access charges. This model, which was initially adopted by a number of European operators as they introduced WAP services over their GSM networks in the late 1990s, was in part to blame for the slow adoption of the service as it typically produced particularly steep charges. You would be billed for accessing WAP services and for the number of minutes you would stay connected—not just the time to download or upload information, but also the time to read and write your email, read the latest stock updates, and so forth.

By also operating as content providers, mobile operators can generate additional revenues, whether in the form of access charges for their content or in the form of advertising fees. As mobile portals, they can charge users additional subscription fees, create additional revenue streams from advertising, or position themselves to capture a portion of the revenues generated by partnering content providers through a variety of revenue-sharing arrangements. More generally, by collecting information about the preferences of their customers, leveraging user position information, or billing the user on behalf of content providers, they hope to position themselves at the center of the m-commerce value chain, and in the process grab a slice of all mobile transactions. Mobile operators such as NTT DoCoMo have shown that business models combining these many sources of revenue with well thought-out partnership arrangements and pricing schemes that encourage broad user adoption can prove extremely successful. In contrast, mobile operators that have been slower in waking up to these new opportunities or have attempted to push high usage fees onto their customers or greedy revenue-sharing models onto potential partners have had to revise their strategies. This was the case with some Nordic operators that were initially taking up to 90 percent of the revenue generated by partnering content providers.

Whether all mobile operators will succeed in quickly redefining themselves seems rather doubtful. For one, they require major economies of scale to recoup their infrastructure and spectrum investments. The result is fierce competition for market share coupled with aggressive expansion plans, which many operators can ill afford. As of mid 2001, European telecom operators such as Deutsche Telekom, France Telecom, KPN, Telefónica, and BT were each operating under debt burdens in excess of $20 billion (in excess of $60 billion in the case of the first two), following major capital outlays for 3G spectrum licenses and various acquisitions. To make matters worse, 2.5G and 3G mobile Internet revenues have taken longer to materialize than originally suggested by analysts and infrastructure equipment providers, due in part to deployment delays, technological glitches, slow user adoption, and unavailability of handsets, to name just a few. The predicament in which many mobile operators find themselves today is not limited to Europe. Many U.S. operators, for instance, are not much better off financially and face additional problems associated with a mosaic of legacy systems (for example, additional roaming and 3G upgrade problems), not to mention delays at the level of the FCC in coming up with a solution that alleviates a severe spectrum shortage.

But beyond these challenges, the biggest threat to mobile operators may end up coming from other players competing for some of the same segments of the value chain, segments where infrastructure ownership does not lend a competitive advantage. Among others, they include:

Virtual mobile network operators (VMNOs). VMNOs do not own spectrum, but instead buy bandwidth from traditional mobile operators for resale to their own customers. An example of VMNO is Virgin Mobile, which relies for its spectrum on mobile operators such as One2One in the UK or SingTel in Singapore. VMNOs are not burdened by heavy debt from the purchase of spectrum licenses and can focus on building a close relationship with the customer through personalization, delivery of high value-added applications and services, and control over billing.

Mobile portals. Most mobile operators have mobile portals, which they hope will help them build a close relationship with their customers. As a result, they are in direct competition with traditional portals that have established a mobile presence as well as a number of new mobile entrants.

Third-party billing providers. Banks, credit card companies, and other billing providers are developing strategies to circumvent the mobile operator's billing relationship with its customers.

These last two categories of players are further discussed later in this chapter. At the end of the day, the real value of mobile operators will lie in their ability to leverage their existing relationship with their customers (for example, billing, position information, and so forth) to assemble a critical mass of partners and deliver highly personalized applications and services. Players that can emulate the early success of NTT DoCoMo can hope to grab a substantial slice of the mobile Internet services market, a market that will likely be worth hundreds of billions of dollars over the coming decade. The others will find themselves squeezed as low-margin, bulk-traffic carriers and will become prime acquisition targets.

Mobile Portals

Mobile portals offer a one-stop shop solution to mobile users by combining mobile Internet and Personal Information Management (PIM) applications such as email, search, calendar, and instant messaging with content and services from a number of providers (see Figure 1.1). Because of the input/output limitations of mobile devices, mobile portals could very well play an even more central role than on the wired Web. To assume this role, they have to develop a close relationship with the customer, acting as central repository for his or her preferences, contextual information such as position, and billing information. Because it may very well hold the key to the m-commerce value chain, the mobile portal market is emerging as a strategic battleground where mobile network operators, traditional Internet portals, and a number of other entrants are vying to position themselves as the m-commerce intermediary of choice. According to *McKinsey Quarterly*, by mid 2001, close to 200 mobile portals had been launched in Europe alone. By 2005, the global mobile portal market could be worth between $5 and $10 billion per year.

Running a mobile portal does not require owning any infrastructure. Operating costs tend to be independent of scale with the exception of sales and marketing. Revenues, on the other hand, are directly tied to the portal's customer base through a combination of monthly subscription fees, advertising revenue, traffic-based revenue sharing arrangements with mobile operators, commissions on transactions, and so forth. The resulting situation is one where a portal's profitability is directly dependent on scale. To compete, mobile portals need to offer their customers a compelling set of services that combine ease of use and personalization with a critical mass of services and applications—all at a reasonable price.

Players in the mobile portal market fall into a number of categories:

Mobile operators. As discussed earlier, repositioning themselves as successful mobile portals is key to the strategy and very survival of many mobile operators today. To succeed, they need to capitalize on their broad customer base and leverage strategic advantages such as their micropayment and authentication infrastructure, their billing relationship with the customer, information about the user's position, and control of the device's first screen. Today's most successful mobile portal, NTT DoCoMo's i-Mode, falls into this first category.

Traditional Internet portals. Traditional Internet portals such as AOL, Yahoo!, MSN, Lycos, or Excite all have established a mobile presence, trying to leverage their existing customer base and broad network of content providers. They hope to develop a more intimate relationship with their users by allowing them to remain in constant contact with their portals, while on the go. This in turn should translate into higher per-user revenues than what these portals have been able to generate so far over the wired Internet. The longer-term strategy of these players is to eventually offer their customers a single, integrated portal solution across all access channels, from PCs to mobile devices, digital TVs, cars, and so forth.

Mass media companies. Mass media companies are also eager to maintain a close relationship with their customers by developing their own mobile portals. In so doing, they hope to gain deeper insight into their customers' preferences and behaviors and position themselves as major intermediaries in the m-commerce value chain. Vivendi Universal is a prime example of a mass media player pursuing this strategy through the development of its Vizzavi portal in partnership with global mobile operator Vodafone.

Device manufacturers. By taking advantage of their large market shares and strong brands, device manufacturers such as Nokia, Motorola, or Palm have shown that they too are anxious to compete for a slice of the mobile portal market. For example, Palm offers its users access to its MyPalm portal (see Figure 2.11), and Nokia has started a Club Nokia portal for its handset users.

Independent mobile portals. The mobile portal market has also attracted a number of new *pure-play* entrants. They include players such as Halebop, djuice, or Mviva. These mobile portals are among the most innovative ones. At the same time, many of them have had to seek backing from major mobile operators or mass media players, or have simply been acquired by larger players. This is the case of iobox,

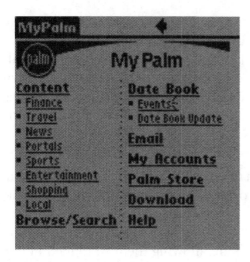

Figure 2.11 MyPalm portal.

Courtesy of Palm.

which was bought in the summer of 2000 by Terra Mobile, a joint venture between Telefónica Móviles and Terra Networks. Others such as Room33 were not as lucky and simply went belly up.

Financial organizations. As discussed in Chapter 1, financial institutions such as Nordea have also set up specialized WAP portals for shopping and banking. In doing so, they are attempting to provide additional convenience to their customers while offering them a micropayment alternative to mobile operators and other mobile portal players.

One question faced by all portals is how to price their content, and whether any of it should be made available to nonpaying users. i-Mode, as we saw, offers some content to its users in return for a basic monthly subscription, while allowing them to subscribe to premium services in return for additional monthly charges. Other portals such as Halebop or Yahoo!Mobile offer all their content for free. As with content providers, there are a number of possible business models, combining different sources of revenue. Mobile operators offering their content for free have to rely on advertising and revenue-sharing arrangements with mobile operators and content providers. By offering their content free of charge, mobile portals can potentially build a larger customer base and generate higher revenue through advertising and revenue sharing. By selectively charging for a subset of their content, they can try to combine the best of both models and have the added flexibility of being able to test the popularity of some of their offerings. As new services become more popular, mobile portals can start charging for them. Clearly, not all models will work for all players. For example, mobile portals run by financial organizations or traditional Internet portals can be subsidized by other operations.

Third-Party Billing and Payment Providers

Mobile transactions such as accessing the latest weather report, requesting a traffic update, or buying a soda tend to result in fairly small charges, making it particularly impractical for many content providers to take care of their own billing. This situation has been compounded by the lack of end-to-end security in early mobile Internet standards such as early versions of WAP or i-Mode, creating a further incentive for content providers to rely on the mobile operator's billing infrastructure—this is further discussed in Chapters 4 and 5. Today, an increasing number of operators are following in DoCoMo's footsteps and offering affiliated content providers to take care of their billing in return for a percentage of their fees (see Figure 1.3). This can be done through the operators' regular billing cycle or through pre-paid phone cards. The result is a situation where mobile operators are challenging the traditional role held by banks and credit card companies. Financial organizations, however, have not been sitting idle. Instead, a number of third-party solutions are emerging that all aim at circumventing the mobile operator's position as default billing and payment provider. An example is the WAP Solo service discussed in Chapter 1, which allows customers to choose from a number of options when making a purchase. A number of variations of this *mobile wallet* concept have recently been launched. They generally involve redirecting the user to a special mobile wallet site, where the user logs in with username and password and selects a particular method of payment; for example, credit card, bank account, debit card (see Figure 1.5). Several initiatives have also been launched to develop standards that will make it easier for third-party payment providers to compete with mobile operators. They include the Mobile electronic Transaction (MeT) initiative, the Mobey forum, the Mobile Payment

Forum and the Global Mobile Commerce Interoperability Group (GMCIG). These initiatives are also discussed in Chapter 5.

Mobile Device Manufacturers

Mobile device manufacturers are key players in the m-commerce value chain. Their design decisions determine the functionality and standards available on mobile platforms:

- How much memory and how fast a CPU will the device have? How large of a screen will it sport? What type of input capability will it support—keypad, voice, pen?

- What operating system will it be running?

- What mobile communication standards will the device support? Will it support GPRS or WCDMA or multiple standards? Will it be Bluetooth-enabled?

- What microbrowser will be pre-installed? Will the device be WAP-enabled? Will it run Java or Qualcomm's "Binary Runtime Environment for Wireless" (BREW), which allows people to write applications that run across different CDMA devices?

- Will it include a slot for a subscriber identity module (SIM), a wireless identity module (WIM), or some other tamper-proof mechanism for supporting certificate-based user authentication and digital signatures? Will it sport a dual slot for debit or credit cards from third-party payment providers?

- What type of location tracking functionality will the device support?

- What applications will be pre-loaded on the device? Will it contain an address book, a calendar, games?

- What other standards will it support? Will it be capable of playing MP3 songs or supporting MPEG4 videostreaming?

- What usage scenarios will its form factors and functionality be more conducive to?

Mobile Internet devices range from simple WAP-enabled mobile phones and wireless PDAs to Internet-enabled wristwatches and a number of new hybrids combining phone and PDA features with possibly those of other devices such as videocameras or MP3 players. Competition in the mobile handset industry has proved extremely fierce. Companies that once seemed to dominate the market, such as Motorola, have seen their market share erode in no time. The emergence of the mobile phone as a consumer appliance is forcing handset manufacturers to operate under much shorter cycles than in the past, and subject to much thinner profit margins than they were accustomed to. In the process, even number-two handset manufacturer Ericsson came to realize that it did not have what it took to continue on its own. In the first half of 2001, the Swedish behemoth announced that it would reorganize all its handset manufacturing operations under a joint 50/50 partnership with Sony. Despite these difficulties, mobile handset manufacturers are in a position of strength in the m-commerce value chain. It has been shown that often customers will shop for a particular handset brand rather than for a particular mobile operator. This position of strength coupled with the control they have over device functionality has allowed them to play a key role in the development of mobile Internet standards. At the same time, as competition in the mobile device market intensifies, manufacturers such as Nokia, Motorola, Samsung, or Ericsson are also trying to diversify and leverage their strong brands to occupy new segments of the m-commerce value chain. Nokia and Ericsson, for example, offer a number of value-added applications and platforms to mobile operators, mobile portals, and content providers. As mentioned earlier, companies such as Nokia, Palm, and Motorola have gone as far as developing their own mobile portals

Wireless Application Service Providers (WASPs)

Because many players across the m-commerce value chain lack the expertise or the resources to develop, maintain, and/or host many of their applications and services on their own, they turn to third-party wireless application service providers (WASPs). As a result, WASPs can be found in nearly all segments of the m-commerce value chain (see Figure 2.12):

Application development. A number of WASPs focus on the development of messaging systems, payment solutions, enterprise applications, or location-based services, to name just a few. They include Oracle Mobile, the mobile products and services division of Oracle, which develops and hosts mobile applications based on the wireless edition of its 9i Web Application Server. Other examples are GoYada with a number of consumer-oriented applications for mobile commerce and messaging, and Webraska, which we discussed in Chapter 1.

Hosting and managing the application. Many WASPs also provide application hosting and management services. They include players such as OracleMobile, 724 Solutions, Aether Systems, Brience, and many more, as well as more specialized companies such as wcities, which develops, hosts, and maintains location-based information on hundreds of cities around the globe for mobile operators, mobile portals, and other m-commerce players.

Developing the wireless bridge. Developing so-called bridging solutions that provide wireless access to consumer and enterprise applications is

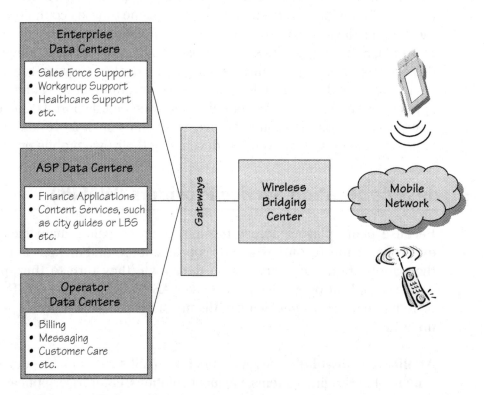

Figure 2.12   WASPs can be found at all levels across the m-commerce value chain. They might just develop and customize applications, or also manage and host them. Many also offer and host wireless bridging solutions.

another area where WASPs such as 724 Solutions, Air2Web, Oracle Mobile, and many more are present.

Hosting and managing the wireless bridge. Many of these same players also offer their customers to host and manage the wireless bridge solutions they develop.

When developing their own applications, mobile operators such as BTCellNet or TeliaMobile can also be viewed as WASPs. BTCellNet, for example, offers corporate customers e-business applications over its GPRS packet-switched network.

In part due to their diversity, WASPs tend to rely on a variety of business models. These models will often combine licensing fees for their applications, one-time integration fees, monthly hosting fees, as well as possibly some revenue-sharing arrangements with mobile operators, mobile portals, or content providers. They might also include advertising fees or referral fees in the case of applications providing recommendations to customers (for example, city guides).

Location Information Brokers

Because of the central role of location-based services in the uptake of the mobile Internet, location information brokers occupy a special place in the m-commerce value chain. They are responsible for determining and updating the user's position and supplying this information to content providers, mobile portals, and other players across the value chain, enabling them to tailor their services to the user's location. The emergence of location-based services is driven by the need to provide users with highly personalized and context-aware services, the demand for location-sensitive applications such as fleet tracking, as well as safety regulations such as E-911 in the United States, which mandates the time phased introduction of emergency location-tracking technology by mobile operators.

Examples of location information brokers include UK-based Cambridge Positioning Systems, whose Enhanced Observed Time of Difference (E-OTD) technology has been adopted by a number of GSM network operators worldwide; or Canadian Cell-Loc, which provides location-tracking information over AMPS and cdma networks, using Time Difference of

Arrival (abbreviated TOA or TDOA) technology—E-OTD, TDOA, and other positioning technologies are discussed in Chapter 7, "Next Generation M-Commerce." Among other key players, one finds infrastructure equipment providers such as Nokia, Ericsson, and Lucent, and a number of niche players such as Allen Telecom, TruePosition, Snap-Track, or Sirf. Many of these solutions are provided in partnership with infrastructure equipment providers or handset and PDA manufacturers, as they often require enhancements to the network and/or the mobile device. As illustrated in Figure 2.13, location information brokers provide positioning information to mobile operators, mobile portals, content providers, and WASPs such as Webraska or Signalsoft that specialize in the delivery of location-based services. These players might in turn make this information available to others, raising some important privacy issues. For example, a mobile portal or mobile operator could, in principle, make the position of its users available to advertisers or possibly other users. Clearly, these scenarios require that the user be able to specify with whom and under which circumstance she feels comfortable having her information disclosed (for example, emergency services, buddy lists, and so forth). Beyond this, the Location Interoperability Forum

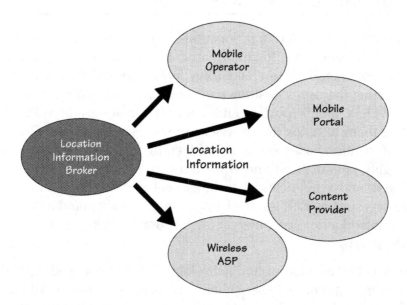

Figure 2.13 Location information brokers supply position information to mobile operators, mobile portals, content providers, and wireless ASPs, enabling them to tailor content to the user's position.

(LIF) is addressing issues of interoperability to allow users to access location-sensitive services as they roam across networks supporting different positioning solutions.

As is the case for many m-commerce players, location information brokers can choose from a number of different business models, some involving simple licensing fees for their solutions, others tied to actual usage levels, or a combination of both.

Concluding Remarks

In this chapter, we took a first look at the overall m-commerce value chain and its many players, from infrastructure providers and handset manufacturers to content providers, mobile operators, mobile portals, and WASPs. As we have seen, the relationships between many of these players can be fairly complex. Together with a number of possible sources of revenue, from usage fees to advertising, referral fees, commissions, and sharing of traffic-based revenue, to name a few, they entail a variety of business models. Providing users with a broad range of compelling and highly personalized, context-sensitive content at a reasonable price is likely to be key to the success of mobile portals, which need large customer bases to survive. M-commerce and the mobile Internet at large are forcing many actors to reconsider their role in the value chain. Mobile operators are becoming mobile portals and mobile payment providers. Mobile device manufacturers are developing their own applications and going as far as starting their own portals. Financial institutions are working on solutions that circumvent the emerging role of mobile operators as payment providers, and so forth.

Many players are coming to m-commerce armed with experience, partnerships, business models, and legacy systems developed over the wired Internet. This is best illustrated by traditional Internet portals such as Yahoo!, which are trying to reposition themselves as global multi-channel portals. While these companies benefit from strong brands, large customer bases, and broad networks of content partners, they also need to revise their business models and solutions if they are to succeed in the mobile world. Ultimately, the lessons they will learn from the mobile Internet, such as providing higher levels of personalization and generating revenues from sources other than advertising, will prove useful over the wired Internet as well.

Finally, because it involves different delivery mechanisms and entails new usage scenarios, services and applications, m-commerce is also creating opportunities for a number of new entrants. They include content providers, location information brokers, Wireless Application Service Providers and many others. Because they generally lack direct access to the customer, need to rely on operators for end-to-end security and cannot justify developing their own billing infrastructure, many of these players are dependent on revenue-sharing arrangements with mobile operators.

PART TWO

The Technologies of M-Commerce

CHAPTER 3

Mobile Communications: The Transition to 3G

Introduction

Chief among the complaints of early mobile Internet users, such as WAP adopters accessing the Internet over GSM, has been their frustration with the slow data rates, long connection setup times, and steep access charges associated with these networks. This merely reflects the fact that 2G communication standards such as GSM were primarily designed with voice applications in mind—Internet services came as an afterthought. One of the major driving forces behind the emergence of m-commerce is the ongoing transition toward faster 2.5G and 3G packet-switched technologies, which we already alluded to in Chapter 1, "M-Commerce: What's the Buzz All About?"

In this chapter, we take a closer look at mobile communications technologies. Here, you will learn about the different standards competing in the marketplace, their adoption levels, and how already deployed standards affect the options available to mobile operators as they upgrade their networks to 2.5G and 3G technologies. This includes a discussion of the cost and spectrum requirements associated with different 3G standards. Deployment of 2.5G and 3G is a high-stakes game for mobile operators, infrastructure providers, handset manufacturers, service providers, and governments. In Europe alone, mobile operators have spent over

$100 billion on 3G licenses and will likely spend somewhere between $50 and $100 billion to deploy the necessary infrastructure. At the same time, the UMTS Forum has claimed that the combined worldwide revenue generated by 3G services over the next decade could approach $1 trillion.

Short of delving into all the gory details of communication technologies—specifications of standards such as GSM contain more than 7,000 pages of documentation—the objective of this chapter is to give you a better understanding of how mobile communication networks differ from fixed wireline networks. In the process, we introduce key architectural elements of these networks and how they interface with wireline ones. We will look at issues such as quality of service (QoS), security, roaming, billing, and more generally, at how the characteristics of mobile communication networks impact the design of successful m-commerce applications and services.

Readers familiar with mobile communications network might want to simply skim through this chapter, possibly just reading the sections dealing with 2.5G and 3G standards.

Mobile Communications: A Quick Primer

Historical Perspective

Strictly speaking, wireless communication dates back to antiquity with, for example, the use of fire signals in ancient Greece being reported by Polybius around 150 BC. Modern-age wireless communication, however, has its origins in the discovery of electromagnetic induction by Michael Faraday in 1831, the theoretical work of James Maxwell who laid the foundation of electromagnetism, and the empirical work of Heinrich Hertz who later demonstrated the wave character of electrical transmission through space. In 1895, Guglielmo Marconi was the first to demonstrate wireless telegraphy over a distance of nearly two kilometers. By 1901, the first wireless signals were already sent over the Atlantic and, by 1907, the first commercial transatlantic communication system was in operation. KDKA, the first commercial radio station in the world, started operating in Pittsburgh in 1920, and within a few years, the first cars equipped with radios started reaching the market. The first commercial system for car-based telephones was set up in 1946 in

St. Louis, Missouri. It was a *push-to-talk* system similar to a CB-radio or the system used by taxi drivers. It relied on a single large transmitter on top of a tall building and only had one channel, used for both sending and receiving—not exactly a solution that could accommodate the hundreds of thousands or even millions of mobile phone users found in many large cities today. A number of early mobile telephony systems were deployed around the world through the 1950s, 1960s, and 1970s with names such as Improved Mobile Telephone System (IMTS) in the United States, or A-Netz in Germany. These systems generally required bulky senders and receivers, which were typically built into cars and supported only very limited roaming functionality. This explains why they never had more than a few thousand users.

The first mobile phone standard to really gain broad user acceptance was the Advanced Mobile Phone System (AMPS), which was first introduced in the United States in 1982—AMPS is also known as International Standard IS-88. The key innovation in AMPS is to divide space in relatively small areas or *cells*, typically 10 to 20 kilometers across, and re-use transmission frequencies in nearby nonadjacent, cells (see Figure 3.1). In so doing, AMPS makes it possible to accommodate a much larger number of users at the same time. The use of relatively small cells also implies that handsets require much less power to communicate with antennas responsible for relaying communication to the wireline network. AMPS phones can operate on 0.6 watts, and car transmitters require about 3 watts in contrast to the 200 watts required by IMTS. The result is much lighter and cheaper handsets than in the past.

Figure 3.1 In AMPS, space is partitioned into small cells, 10 to 20 kilometers across. Cells are organized in clusters, with groups of frequencies f_i being reused in nonadjacent cells. Here, we show a pattern with clusters of seven cells.

Basic Architecture

Figure 3.2 illustrates some of the key architectural elements of a generic mobile phone system. They include base stations (or base transceiver stations—BTSs) located at the center of each cell and responsible for allocating communication channels to mobile stations—another term for mobile devices—and relaying communication to the core wireline phone networks via intermediate base station controllers (BSCs). As users—and their mobile stations—move from one cell to the next, BSCs are responsible for handing over responsibility for that mobile station from one base station to the next. Each BSC is responsible for all BTSs in a particular geographical area. At a higher level within the architecture, mobile switching centers (MSCs) perform additional book-keeping tasks, making it possible to efficiently find, authenticate, and bill users as they roam from one geographical area to another. This will be further detailed as we discuss 2G networks.

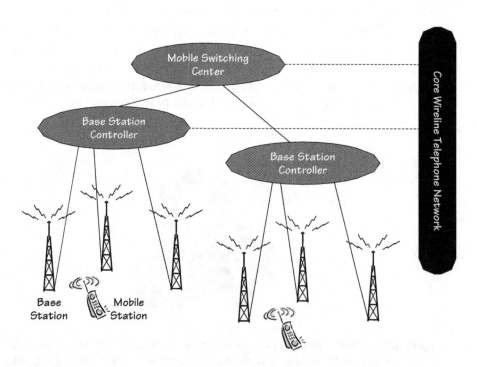

Figure 3.2   Basic architectural elements of a typical mobile phone system.

What Is So Special about Mobile Communication?

Mobile wireless communication introduces a number of challenges not found—or at least not as severe—in fixed wireline networks, leading to different architectural features and protocols. They include:

Unpredictable medium. Radio transmission cannot be shielded as is the case with a coaxial cable. Weather conditions or nearby electrical engines can cause severe interference, resulting in high data loss rates or high bit error rates. As users move around, the properties of the communication medium change: a building might prevent transmission or the user might find herself too far from any base station. This also provides for some interesting challenges when it comes to adapting Internet protocols (TCP/IP) to run on mobile communication networks. Additional challenges arise when implementing m-commerce solutions over these networks. Imagine a user initiating a transaction from her mobile phone as she enters a tunnel and loses her signal. Special care must be taken to ensure that the transaction is properly executed or, at least, that the user is informed of its status, so that she can restart the transaction if it was not completed the first time.

Low bandwidth. While they are increasing with the introduction of 2.5G and 3G technologies, data rates associated with mobile communication networks remain low in comparison with wireline networks.

Shared medium and limited spectrum. Wireless communication always takes place through a shared medium for which multiple users often have to compete. As the number of users in a particular cell increases, the communication channels they have each been allocated might start interfering with one another. At some point, it might be best to no longer accept any additional users. Cells in cdma networks actually adapt their coverage to the number of users and essentially shrink as the number of users increases, relying on adjacent cells to pick up the extra load. In general, because spectrum is limited, it is regulated, with the exception of a few frequency bands, and clever multiplexing protocols have to be devised to efficiently share available spectrum among competing users. As we will see, there are several basic multiplexing schemes available, and each standard has its own peculiar way of combining them. Beyond this, quality of service (QoS) considerations—namely, ensuring delivery of a consistent and predictable service to the user in terms of delays, bit rates, and other key parameters—play a central role in the design of mobile communication systems.

Security challenges. Mobile communication is prone to a number of security challenges not found in traditional wireline networks. For one, mobile devices are much easier to steal than desktop computers, and are also more likely to be forgotten by their users. Beyond this, the air interface over which communication takes place is also more prone to eavesdropping. Analog communication networks such as AMPS are totally insecure. A number of incidents have been reported with AMPS where people with scanners would eavesdrop on conversations, steal credit card numbers, or monitor the network's control channels to steal telephone numbers, which they would later resell in one form or another. As we will see, more recent mobile communication standards have shown that it is possible to address these problems with special encryption, authentication, and other security mechanisms.

Mobility challenges. As users move from one cell to another, mobile communication networks need to ensure continuity of service. This involves transferring responsibility for the user from one base station to another, from one base station controller to another, and, possibly, from one network standard to another and from one mobile operator to another. In the process, solutions need to be devised to provide for as seamless and secure a transition as possible, rerouting voice and data communication as needed. This also involves making sure that all charges get posted to the user's account, while taking care of splitting revenue among all parties involved in the delivery of the service.

As designers of mobile communication networks and mobile Internet services attempt to address these challenges, they need to also take into account the many limitations of mobile devices: their limited processing power, battery life, memory, and input/output functionality. QoS issues also require distinguishing between the requirements of different types of services. Having a phone conversation, accessing email, playing an interactive videogame, or purchasing stock involves very different QoS requirements; for example, different delay requirements, different data integrity requirements, different bit rate requirements, and so forth.

Basic Multiplexing Schemes

One of the distinguishing features of a mobile communication system has to do with the way in which it separates users from one another, or what people refer to as their *multiplexing* protocols. Multiplexing enables users to share a communication medium with minimum or no interference, and generally aims at maximizing the number of users that can be accommodated at the same time. As we just saw, AMPS relies, among other things, on separating users based on their particular location; namely, the cell in which they currently are. Each cell has a number of 30-kHz-wide communication channels allocated to it, making it possible for multiple users to use their phones at the same time. As a user finishes a phone conversation, the communication channel she was using is freed and made available to other potential users. The result is a system that separates users based on location, frequency, and time—the time when the phone conversation takes place. As it turns out, these represent three of the four basic multiplexing schemes used to separate users in mobile communication networks:

Space Division Multiplexing (SDM) involves separating users based on their location (see Figure 3.3). As we have already seen, this can be done using cells and allowing multiple users in different cells to possibly use channels with the same frequencies at the same time. If the cells are sufficiently far from one another, the signal in one cell will not interfere with the one in the other cell.

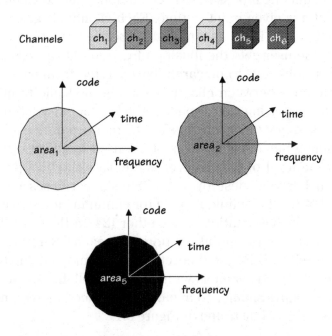

Figure 3.3 Space Division Multiplexing.

Frequency Division Multiplexing (FDM) partitions the available spectrum in nonoverlapping frequency bands, allowing multiple users, all possibly at the same location, to be concurrently allocated channels with different frequency bands (see Figure 3.4). This scheme is commonly used for radio stations broadcasting in the same area. Each station is allocated a channel that it can use 24 hours a day. Clearly, in the case of mobile communication systems where users might only need half an hour of airtime each day, FDM is not a very efficient solution if used in isolation.

Time Division Multiplexing (TDM) allows multiple users to share a given frequency band as long as they use it at different times (see Figure 3.5). This requires some form of coordination among mobile stations, typically via the base station. As we have already seen, TDM can be combined with FDM (and SDM). This is done in a straightforward fashion in AMPS. More advanced standards such as GSM use more sophisticated variations of TDM, where multiple users can share the same frequency band during overlapping time intervals. As illustrated in Figure 3.6, this is done by partitioning the frequency band into very short time slots (for example, 577 microseconds in the case of GSM), which are allocated in a *round-robin* fashion. Specifically, in GSM, each frequency band is partitioned into time frames consisting of eight time slots, making it possible to accommodate up to eight users (or channels) at any given point in time. Each channel is guaranteed the same n-th time slot in every time frame. Because the time frames are very short, the user gets the illusion of continuous access to the network. Times slots have to be carefully separated from each other to avoid interference between channels sharing the same frequency band. This is done by using guard spaces. The format of each time slot also provides for special bits (*tail* bits) to further separate each slot. Additional training bits in each time slot are used to adjust different parameters at the level of the receiver and ensure the best possible communication. In its simplest form, the GSM standard uses a total of 248 channels, 124 for the uplink, namely for communication from the mobile station to the base station and another 124 for the downlink—from the base station to the mobile station. With each GSM frame partitioned in 8 time slots, GSM can theoretically accommodate up to 992 (in other words, 124×8) channels within a given cell. In practice, the number is significantly smaller to avoid interference between neighboring cells, as already illustrated in Figure 3.1.

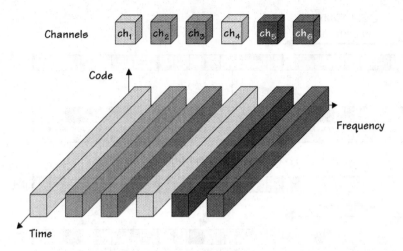

Figure 3.4   Frequency Division Multiplexing.

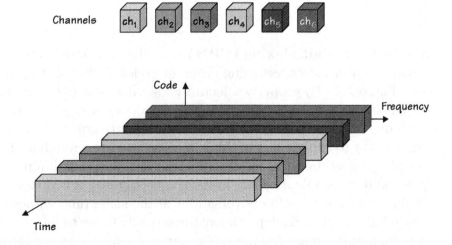

Figure 3.5   Time Division Multiplexing.

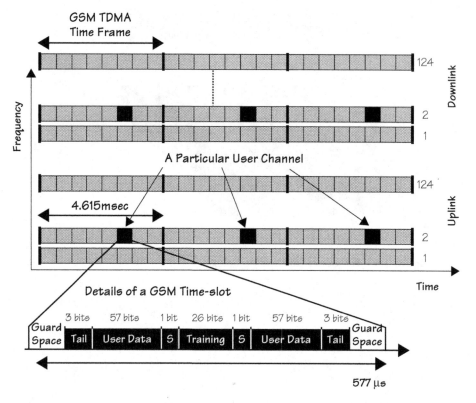

Figure 3.6 Combining TDM and FDM in GSM. Each GSM TDMA frame is partitioned into eight time slots of 577 microseconds (μs). A user channel is allocated one time slot in each consecutive TDMA frame.

Code Division Multiplexing (CDM) is a relatively new multiplexing scheme that involves assigning different codes to different channels (see Figure 3.7). By properly selecting codes, it is possible for multiple channels to coexist in space, time, and frequency without interfering with one another. Without going into all the nitty-gritty of CDM, the easiest way to explain it is to think of a party where a number of people all speak at the same time. If different people speak different languages (for example, English, French, and Chinese), it is possible for different groups to hold conversations at the same time without too much interference. Each participant essentially tunes to the sound of a particular language and the other languages appear as background noise. If the languages spoken by some groups are very close (for example, Danish and Norwegian) or the same, people in these groups will find it much more difficult to follow what is being said. Similarly, in CDM, codes are selected to minimize interference, making it possible for multiple pairs of senders and receivers (or channels) to all communicate at the same time in the same frequency band. As it turns out, the number of non-interfering codes that one can come up with in a given frequency band is quite large compared to the number of non-interfering channels one could support using FDM. CDM also has built-in security in the sense that if the code used by a given sender/receiver pair is kept secret, no one else can follow what they are saying or understand the data they are exchanging. CDM is, however, significantly more complex than the other three multiplexing schemes. Beyond coordinating code allocation, it also requires receivers capable of separating the channel with user data from background noise. This in turn requires careful power control and synchronization between the mobile station and base station. In contrast to TDM, however, CDM does not require any synchronization between different channels. CDM is used in the cdmaOne standard and in two out of the three 3G standards, namely cdma2000 and WCDMA/UMTS.

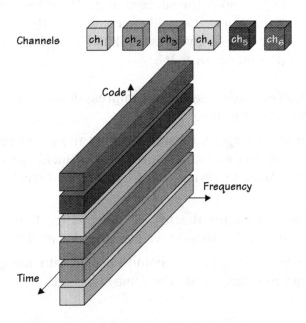

Figure 3.7 Code Division Multiplexing.

Separating Uplink and Downlink Traffic

Besides separating users from each other, mobile communication systems also need to worry about separating uplink and downlink traffic; namely, traffic respectively from and to the mobile station. Here, one essentially distinguishes between two approaches:

- **Time Division Duplex** (TDD) assigns different time slots to the uplink and downlink, while using the same frequency band for both. Bluetooth, which is discussed later in this chapter, is an example of a standard relying on TDD.
- **Frequency Division Duplex** (FDD), on the other hand, allocates separate frequency bands to the downlink and uplink, as already illustrated in Figure 3.6 in the case of the GSM standard.

The 2G Landscape

AMPS was the first mobile phone system to be adopted on a broad scale. It is an analog standard, where essentially an analog image of the sound is transmitted over the air interface. Such standards today are referred to as *first generation* (1G). Despite the success of AMPS in the United States and other parts of the world, it quickly became clear that analog standards would eventually have to give way to digital ones. Digital systems offer a number of advantages over 1G:

- By digitizing voice, it becomes possible to compress data and achieve a better use of available spectrum.
- Digital signals can be encrypted, thereby offering the potential for greater security—as discussed earlier, a number of incidents have been reported with AMPS, involving eavesdropping and stealing of phone numbers.
- Digital transmission allows for the use of error correction techniques, which make it possible to improve transmission quality.
- With digital voice transmission, it becomes possible to integrate voice, fax, and data transmission into the same standard.

Digital mobile communication systems were introduced in the early 1990s. These systems are generically referred to as second-generation (2G) networks.

In the United States, where AMPS had already gained a sizable customer base, a key concern of infrastructure equipment providers and mobile operators, as they started development of 2G systems, was backward compatibility with 1G networks. They wanted a technology that would make it possible for users to rely on 2G communication networks, where available, while being able to take advantage of already deployed AMPS infrastructure elsewhere. The result was a 2G standard referred to as Digital AMPS, or D-AMPS, which is now more generally referred to as TDMA (or International Standard IS-136). In densely populated areas, where TDMA technology has been deployed, users connect via a 2G network, while they can fall back on the 1G infrastructure when roaming in more sparsely populated areas where the operator has not yet made the necessary investments.

In Europe, on the other hand, nearly every country had developed its own analog system, each relying on standards incompatible with those of other national networks. The resulting situation was untenable when it came to achieving economies of scale or supporting roaming across national borders, a particularly important issue in a continent that was moving toward a single, integrated market. Instead of attempting to be backward compatible with 1G networks—an impossible proposition in any case, given the number of existing standards—governments and operators set out to develop a brand new standard that would support roaming across the entire continent. This standard, known as Global System for Mobile Communication (GSM), was essentially intended as a wireless counterpart to the wireline Integrated Services Digital Network (ISDN). GSM was first introduced in Europe in the early 1990s and now supports data rates of 9.6 and 14.4 kbps. Thanks to its broad adoption across Europe, it quickly evolved to become the dominant 2G standard with 564 million users worldwide by July 2001, including 324 million in Europe alone. This amounts to about 70 percent of the global 2G population, which, as of July 2001, was estimated at about 800 million (see Figure 3.8).

Over time, several other 2G standards have found their way into the market. Of particular interest is the cdmaOne (IS-95) standard introduced by Qualcomm in the mid 1990s. This technology, which relies on more efficient code division multiplexing technology, was first deployed by

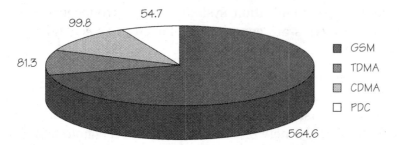

Figure 3.8 Millions of 2G subscribers as of July 2001. This figure ignores standards such as PHS or iDEN, which each have a few million users.

Source: EMC World Cellular Database.

Hutchinson Telecom and has gained broad acceptance in the United States, Korea, Japan, and several other countries in Asia and the Americas. As of July 2001, close to 100 million mobile phone users were relying on the CDMA standard. In Japan, a system known as Personal Digital Cellular (PDC) emerged as the dominant 2G standard. Like GSM and TDMA, it relies on Time Division Multiplexing technology. By the summer of 2001, PDC counted over 50 million users, all in Japan.

A Closer Look at GSM

Work on GSM started in Europe in 1982 with the first set of specifications published in 1990 by the European Telecommunication Standards Institute (ETSI). Commercial GSM networks became operational in Europe starting in 1991. Today, GSM is in use in over 170 countries worldwide. The first GSM networks to be deployed operated in the 900-MHz frequency band and are sometimes referred to as GSM900. More recent versions of the standards also operate at 1800 MHz and at 1900 MHz—PCS frequency band in the United States.

Key features of the GSM standard include:

- **Digital voice transmission**, where the sound of the speaker's voice is sampled and filtered to obtain a digital signal, which is transmitted over the wireless network. This signal is later used by the receiving party to reconstruct the user's voice.

- **Support for global roaming**, making it possible for a user to remain connected while moving from one network operator to another and from one country to another. This is further detailed later in the chapter as we look at some of the main architectural elements of GSM networks.

- **Authentication infrastructure** based on the use of subscriber identity modules (SIM). SIM modules are smart cards that are inserted in GSM phones. They store personal user information, including the user's telephone number and personal telephone book, a PIN code to unlock the mobile device, a personal authentication key used by the network to authenticate the user, and a list of services to which the user subscribes. By simply replacing the SIM card in his phone, a user can change his telephone number and have charges posted to a different account; for example, a corporate account while working for his company, and a personal account when leaving the office and placing personal calls. When buying a new mobile phone, a user can keep the same telephone number and billing information by just inserting his existing SIM card in the new handset.

- **Encryption on the wireless link**, which provides for protection against eavesdropping.

- **Efficient interoperation with ISDN networks** through mobile switching centers (MSCs).

- **Voice mail, call waiting, and a number of other services such as the Short Message Service (SMS)**. SMS relies on unused capacity in the signaling channels to support transmission of messages of up to 160 characters. Use of SMS messages has exploded worldwide. As of late 2001, close to 20 billion SMS messages are sent each month. SMS messages have become a substantial source of revenue for many GSM operators, reflecting in part the emergence of an instant messaging culture among teenagers and young professionals, as well as broad adoption by a number of early mobile service providers to deliver news updates, traffic information, or promotional messages. SMS is now also available in other 2G standards such as CDMA or TDMA.

- **Circuit-switched data services** with peak rates of 9.6 kbps or 14.4 kbps, depending on the specific network.

Figure 3.9 provides an overview of the GSM architecture. A GSM system consists of three component subsystems:

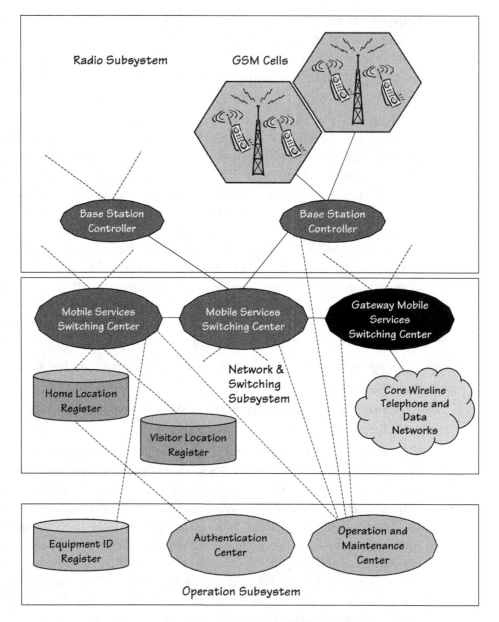

Figure 3.9 GSM system architecture: main functional elements.

- **The Radio Subsystem** includes all the elements directly responsible for radio transmission; namely, the mobile stations, base stations, and base station controllers. This is similar to the generic architecture discussed previously with the exception that communication is now encrypted. An encryption key, which is stored both in the network and on the user's mobile station, is used to both encrypt and decrypt communication at both ends of the wireless link. This key is dynamically generated each time a new connection is set up, by combining the user's authentication key—also stored on the mobile station and in the network—and a random number exchanged between the network and the mobile station. The authentication and encryption keys themselves are never communicated over the wireless link—this is further detailed in Chapter 5, "Mobile Security and Payment".

- **The Network and Switching Subsystem** is the heart of the GSM system, connecting the wireless network with the core telephone and data networks (for example, ISDN, PSTN, X.25 networks) via Gateway Mobile Services Switching Centers (MSCs). It is also responsible for handovers between Base Station Controllers, supports worldwide localization of users as well as charging, accounting and roaming as users connect via the networks of different mobile operators, while possibly moving from one country to another.

The Network and Switching Subsystem also contains two key databases, namely:

- The **Home Location Register (HLR)**: This database serves as a repository for all user information, including the user's telephone number and his authentication key—a copy of which is also maintained on his SIM module. Each mobile operator maintains one or more HLRs with information about each of its users typically stored in only one HLR. The HLR also maintains dynamic information about the user such as the local area where his mobile station is currently located. Local area information stored in the user's HLR allows to keep track of his location across GSM networks worldwide. The HLR also helps maintain important billing and accounting information, as users roam across different GSM networks.

- The **Visitor Location Register (VLR)**: Each MSC has a VLR database. This database helps store information about all users currently in the local area associated with the MSC. As new users move into an MSC's local area, information from their HLR is copied into the VLR so that it can be accessed more easily.

- **The Operation Subsystem** is reponsible for all network operation and maintenance activities. In particular, it includes:

- **Operation and Maintenance Center**, which takes care of functions such as traffic monitoring, monitoring the status of different network entities, accounting, and billing.

- **Authentication Center**, which may actually be collocated with the HLR, is responsible for running authentication algorithms, using keys stored in the HLR.

- **Equipment Identity Register** stores identification numbers for all devices registered with the network. It also maintains a list of mobile stations reported stolen or lost. This is particularly important, given that anyone can use a GSM phone by simply inserting a new SIM module in it.

Despite a number of innovations, GSM suffers from several major problems when it comes to supporting mobile Internet services. Key among them is its reliance on circuit-switched technology. Circuit-switched systems involve long setup times each time a new connection is established. For example, if you drive into a tunnel and lose your connection, it will often take 15 to 30 seconds to reconnect. Circuit-switched technology is also inefficient in its use of available bandwidth, as it requires maintaining a communication channel for the entire duration of a session—not just the time to download your email or the list of movies playing at the theater, but also the time to compose email replies or the time to go over the movies before reserving a seat at the theater. This in turn translates into particularly high usage charges. Low data rates, lack of integration with Internet protocols, and patchy security are yet other areas in which GSM falls short.

A Word about Roaming and Billing

Roaming refers to the capability of a mobile user to access mobile services from outside of his operator's *home* area. This happens, for example, when a mobile user with a plan from, say, Orange in Belgium travels to Spain and attempts to connect via Telefonica Moviles. For roaming to occur, the host network operator—namely Telefonica—needs to have a roaming arrangement with the user's home operator, Orange-Belgium. Roaming agreements among operators specify methods of invoicing for usage of the host operator's network, terms and conditions of payment between the roaming partners, fraud responsibility provisions, and so forth. In places such as Europe with dozens of operators all using the same GSM standard and hundreds of thousands of mobile phone users crossing borders every day, the number of roaming agreements has sky-rocketed. The GSM Association estimates that there are 20,000 roaming agreements among GSM operators worldwide. Because of the complexity of managing such a large number of roaming arrangements, operators are increasingly relying on third-party ASPs for their billing and roaming needs. Third-party solutions today help host operators create *Roaming Call Records*. These records are used to invoice the user's home operator, which in turn sends invoices to its customer, through its regular billing cycle, and collects the charges. Data clearinghouses are responsible for reconciling roaming charges of operators by issuing periodic reports (say, every week or month). In the simplest case, operators will use these records to invoice each other on a regular basis. Increasingly, however, because of the cash flow implications of sending and receiving payments to and from a large number of roaming partners, they prefer the clearinghouses to perform a multilateral net settlement. In this case, the clearinghouse pools together the accounts of all roaming partners and computes how much each company owes to (or is owed by) the pool.

Transition Toward 3G

It takes about 10 years to develop and deploy a new generation of mobile telecommunication standards. Just around the time when 2G networks such as GSM were being deployed, work started on the development of 3G systems. Major motivations behind the development of these new standards included the expected demand for higher data rates, the need to provide for a better interface with the Internet, which among other things involves moving away from circuit-switched technology, and the desire to support a broader range of QoS options, as required by a variety of emerging applications and services. Global roaming, which creates added user convenience and opens the door to new economies of scale, was another important element in the original 3G vision. However, with several hundred million 2G mobile phone users, backward compatibility considerations and the desire to provide for an early, yet gradual transition to faster 3G services have provided for a situation where different operators have chosen different migration paths, many involving intermediate 2.5G technologies (see Figure 1.9). Not all of these paths lead to the same technology. As discussed earlier, there are essentially three emerging 3G standards: WCDMA/UMTS, cdma2000, and EDGE. The migration paths selected by different mobile operators reflect the 2G technologies they already have in place, the costs of upgrading their networks, and possibly acquiring additional spectrum, where available. Initial efforts aimed at developing 3G standards have been coordinated by the International Communication Union (ITU) under its International Mobile Telecommunications 2000 (IMT-2000) program. Today, the brunt of standardization activities is taking place under the 3GPP partnership for EDGE and WCDMA/UMTS, and under 3GPP2 for cdma2000 technologies. Next, we take a closer look at major migration paths for each family of 2G standards.

GSM Migration

As we saw in Figure 3.6, GSM relies on TDMA frames that are divided into eight time slots, each allocated to a different communication channel. A straightforward way to increase peak data rates is to allow a given channel to be allocated more than one time slot in each TDMA frame. This idea is the basis for a simple enhancement to GSM known as high-speed circuit-switched data (HSCSD), which makes it possible, at least in theory, to increase GSM data rates by a factor of 8—in practice, peak rates are closer to 40 kbps. HSCSD only requires software upgrades to the GSM network, and hence is very cheap to deploy. However, as its name indicates, it remains a circuit-switched technology. Figure 3.10 outlines the migration paths available to GSM operators as they upgrade their networks to support faster data rates and packet-switched communication.

GPRS

For most GSM operators, the first step toward supporting *always-on*, packet-switched communication involves deploying General Packet Radio Services (GPRS) technology. The way in which GPRS increases the data rate associated with the GSM air interface is similar to the one used in HSCSD: It allows channels to be allocated more than one time slot per TDMA frame. However, rather than requiring a fixed allocation scheme, GPRS allows the number of time slots allocated to a channel to vary over time to flexibly adjust to actual traffic. The result is a much more efficient utilization of the available bandwidth.

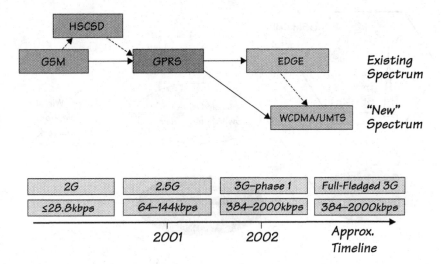

Figure 3.10 GSM migration paths.

GPRS is a 2.5G overlay technology, which means it can easily be deployed on top of GSM networks. As Figure 3.11 illustrates, deploying GPRS on top of a GSM infrastructure essentially involves:

- Changes at the level of the air interface in the form of software upgrades to base stations, and the distribution of new handsets to those who want to access the new packet-switched services—others can continue using their GSM handsets.

- Hardware and software upgrades to base station controllers with, in particular, the introduction of a packet control unit (PCU) to separate circuit-switched and packet-switched traffic.

- Deployment of a separate core network to manage packet-switched data.

Because transitioning from GSM to GPRS does not require any hardware modifications to base stations—imagine the cost of sending technicians to upgrade or replace thousands of base stations located on rooftops and hills—it amounts to a relatively cheap upgrade.

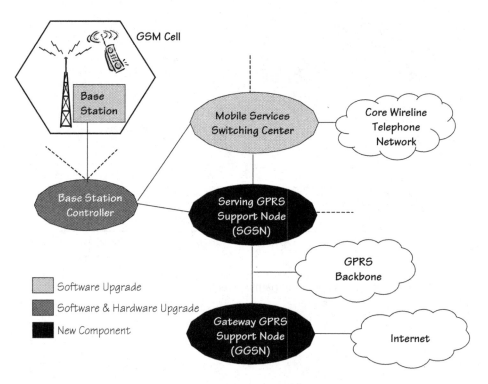

Figure 3.11 GPRS is an overlay solution that primarily involves software upgrades along with the addition of new nodes responsible for managing data traffic.

In the GPRS architecture, each base station controller is connected to a serving GPRS support node (SGSN). For each mobile station under its responsibility, the SGSN keeps track of the base station controller through which it is currently connected. It takes care of forwarding data to the proper BSC, as the user roams from one BSC area to another. Essentially, the SGSN acts as a router that buffers and forwards packets to the mobile station. When a user moves into an area under the responsibility of another SGSN, packets buffered in the old SGSN are discarded and copies are sent to the new one. A separate gateway GPRS support node (GGSN) acts as interface between the GPRS network and the Internet, giving to the outside world the impression that each mobile station operates as a regular Internet node. In practice, the GGSN implements what is known as a *tunneling protocol*—the GPRS tunneling protocol—where packets intended for a particular mobile station are encapsulated into a second packet and redirected to the proper SGSN. Upon reception, the SGSN retrieves the original packet and forwards it to the mobile station via the base station controller currently responsible for it. The network connecting the SGSN and GGSN nodes to each other and to other key components such as the billing gateway used by operators to keep track of user traffic and maintain billing information is referred to as the GPRS backbone. It is a regular IP network with routers and firewalls.

In comparison with GSM, GPRS offers a number of advantages:

- Always-on functionality
- More efficient capacity utilization
- More flexible billing options
- Faster data rates

Specifically, in GSM, to remain connected, the user needs to maintain an open channel by reserving a time slot in a TDMA frame, whether or not that time slot is actually used. Because this time slot is unavailable for use by anyone else, the mobile operator has no choice but to charge the user for the corresponding airtime. In addition, when the user reconnects, say, to check his email or the latest stock news, a new connection setup sequence has to be initiated, which can take 15 to 30 seconds. In contrast, with GPRS, because bandwidth is only reserved as needed, a user can maintain an open connection without consuming any capacity. The result is an always-on environment, where the user can remain connected, while only getting billed for the actual capacity he consumes. Forget also about the 15 to 30 seconds it takes to reconnect when you emerge from a tunnel. *Always-on* means that you are instantly reconnected. On average, this also makes it possible for the operator to accommodate a greater number of users. Moreover, because GPRS users can be dynamically allocated multiple slots as needed, they can achieve much higher data rates. In theory, GPRS is capable of supporting peak data rates in excess of 100 kbps, although actual rates have been reported to be closer to 40 kbps—still a significant improvement over GSM. Another interesting aspect of GPRS is its support for a variety of QoS profiles. In practice, however, limitations at the level of the packet control unit (PCU) are such that most of the time, GPRS QoS will be *best effort*.

It has been estimated that the total cost of upgrading to GPRS for a GSM operator with a user base of 1 million is approximately $10 million, an investment most operators can hope to recoup fairly quickly. In June

2000, BT CellNet was the first GSM operator to upgrade to GPRS. Since then, many operators have followed suit.

Beyond GPRS

Beyond GPRS, operators have essentially two options to move to 3G: EDGE and WCDMA/UMTS (see Figure 3.10). Both technologies use the underlying GPRS infrastructure as their core packet data network.

Enhanced Data Rates for Global Evolution

Coming from GPRS, EDGE only involves the deployment of new modulation technology; namely, technology to encode data for transmission over the air interface. Instead of relying on the binary coding scheme used by GSM and GPRS, EDGE introduces a more efficient modulation technique capable of encoding 3 bits of information at once. You can also think of EDGE as a way of squeezing more data into each GSM/GPRS time slot. As a result, it is capable of supporting peak data rates of up to 384 kbps, while only requiring minor upgrades to the GPRS infrastructure—only at the level of the air interface. In addition, because it can operate in the same frequency band as GSM and GPRS, EDGE does not require new spectrum licenses. For these reasons, it has come to be viewed by a number of operators as a significantly cheaper alternative to the more complex WCDMA/UMTS standard. Some operators have also indicated that they plan to first deploy EDGE technology, and later selectively deploy WCDMA/UMTS in densely populated areas, where they are more likely to recoup their investment. When roaming around the countryside, their users will simply be switched to their EDGE or GPRS networks. As we will see, a variation of EDGE has also been developed to support similar upgrades of TDMA networks.

Like other 3G standards, EDGE supports a number of QoS profiles. These profiles, which are similar to those supported by WCDM/UMTS, are discussed in the next section.

WCDMA/UMTS

WCDMA/UMTS is the 3G standard jointly developed by Europeans and Japanese. It was first deployed in Japan by DoCoMo in October 2001 and is in the process of being introduced by a number of other operators worldwide. WCDMA/UMTS relies on the same core data network as

GPRS and EDGE do. It introduces two completely new wideband CDMA radio interfaces: one based on frequency division duplex (FDD) for outdoor use, and one for indoor use based on time division duplex (TDD), which we will not discuss. The new radio interface requires new spectrum, new base stations, and new base station controllers—referred to as radio network controllers (RNCs). Because deployment of this new air interface is rather expensive, most operators will start by selectively deploying WCDMA/UMTS in urban areas, where they can easily recover their investment, while relying on GSM, GPRS, and EDGE elsewhere (see Figure 3.12). Accordingly, most WCDMA/UMTS handsets are expected to be compatible with the GSM/GPRS air interface. Some handsets will also likely be compatible with the cdma2000 air interface providing for truly global roaming—to the extent that operators can work out the required roaming arrangements. WCDMA/UMTS will eventually support peak data rates of 2 Mbps, although, as always, actual rates will generally be significantly lower—perhaps around 400 kbps. As with other CDMA standards, WCDMA/UMTS data rates vary with the distance between the mobile handset and the base station, with smaller distances generally allowing for faster transmission. As the distance increases, power levels need to be increased to maintain the same data rate. Clearly, this can only be done up to a limit—handsets have only so much power. The RNCs implement mechanisms that attempt to maintain data rates compatible with the QoS profiles of each user.

Like GPRS and EDGE, UMTS/WCDMA allows for a number of QoS profiles. However, in contrast to GPRS where the packet control unit limits the ability of the network to actually satisfy QoS requirements, UMTS provides support for genuine end-to-end QoS management, thanks to its radio network controllers. In particular, it distinguishes between the following traffic classes:

- **Conversational.** This class is for applications such as voice and intensive games that require low delays and the preservation of time relation between packets.

- **Streaming.** As its name indicates, this class is for multimedia streaming applications, where preserving the time relation between packets is critical, although actual delays are somewhat less important.

- **Interactive.** This is for Web browsing and most m-commerce services and applications such as purchasing movie tickets or looking for a nearby restaurant. This class guarantees a decent response time

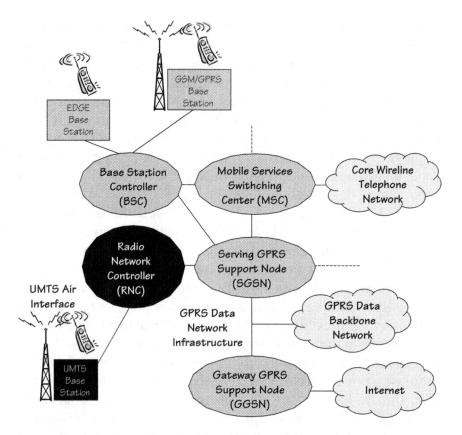

Figure 3.12 WCDMA/UMTS involves the deployment of new base stations and the introduction of radio network controllers, but shares the GPRS data network infrastructure.

and preserves data integrity. Because it is not as demanding in terms of delays as the conversational class, it will also typically come with lower charges.

■ **Background.** This is the most basic class and will likely involve significant discounts in comparison to the others. It preserves data integrity and provides best effort QoS. In other words, delays might vary widely and be as long as a minute or more when network load is particularly high. This class is particularly well suited for background applications such as synchronizing your calendar or downloading a new game.

Beyond these classes, tens of parameters can be specified, such as the maximum required bit rate, a guaranteed minimum bit rate, or a maximum transfer delay. These parameters and traffic classes make it possible for operators to offer a variety of plans to their customers and to differentiate between the requirements of individual types of services and applications. Mobile application developers and content providers will, however, have to figure how to best set up these many parameters so that their applications and services degrade gracefully as users move into areas that are highly congested or where coverage is limited to EDGE or GPRS.

TDMA Migration

TDMA (IS-136) is mainly found in the United States, where it is used by operators such as AT&T Wireless or CellularOne, and in some Latin American countries. Like GSM, it relies on time division multiplexing technology. Since the early 1990s, TDMA networks have been enhanced to support packet-switched data services, using cellular digital packet data (CDPD) technology, also known as International Standard IS-54. CDPD uses the very same 30-kHz channels and essentially the same network infrastructure as AMPS and TDMA. It only requires a fairly minor upgrade to AMPS/TDMA base stations.

CDPD, which includes an Internet Protocol (IP) stack, can run Internet applications and supports peak data rates of 19.2 kbps—closer to 10 kbps in practice. While it has been in use since the early 1990s, CDPD only gained a sizeable consumer base after AT&T Wireless used it to introduce its PocketNet mobile portal in 1996.

TDMA mobile operators have several options, as they develop plans to upgrade their networks (see Figure 3.13). In particular, a variation of EDGE, known as Compact EDGE, has been developed specifically to accommodate migration from TDMA networks. Operators can choose to migrate directly to Compact EDGE, although many, such as AT&T Wireless, have opted to first deploy GSM/GPRS networks and then migrate to Classic EDGE—the name given to the variation of EDGE built using the 200-kHz GSM/GPRS channels in contrast to the 30-kHz TDMA channels used by Compact EDGE. Even though moving to Classic EDGE requires freeing up some bandwidth, it offers the prospect of much broader roaming than Compact EDGE. Also, once they have GPRS in place, operators can very well opt to upgrade to WCDMA/UMTS, which, by and large, is expected to be the dominant 3G standard. The result is again a broader range of roaming options for their customers. This is the route selected by AT&T Wireless.

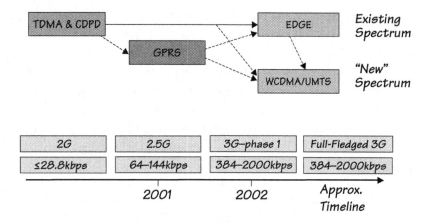

Figure 3.13 Possible TDMA migration paths.

PDC Migration

PDC, the dominant standard in Japan, also relies on a variation of time division multiplexing technology. Since 1997, it has been enhanced with a packet-switched overlay solution known as PDC-P. PDC-P is the network on which NTT DoCoMo launched its i-Mode service. It initially operated at 9.6 kbps, although downlink transmission rates have recently been increased to 28.8 kbps.

Early on, Japanese companies realized that their reliance on a standard used only in their country was a major limitation. To avoid a repeat of this situation, they partnered with Europeans under the 3GPP forum to develop what has become the UMTS/WCDMA standard. In October 2001, NTT DoCoMo was the first operator to offer commercial WCDMA services to its customers with peak downlink data rates of 384 kbps. Overall infrastructure deployment costs of WCDMA by NTT DoCoMo have been estimated at around $10 billion, spread over several years. Initial deployment in October 2001 was limited to central Tokyo, with coverage in other major metropolitan areas expected in the spring of 2002.

CdmaOne Migration

cdmaOne (IS-95a) is the only 2G standard based on code division multiplexing technology. It relies on frequency bands of 1.25 MHz—much wider than the 200-kHz bands used by GSM or the 30-kHz bands used by

AMPS and TDMA. Generally speaking, cdmaOne's migration to 3G is similar to that of GSM with the introduction of faster data rates in a variation of the standard known as IS-95b, as well as the introduction of packet-switched functionality with a 2.5G technology known as CDMA2000 1X (see Figure 3.14). However, there is no strict one-to-one correspondence between the migration of GSM and that of cdmaOne. This is in part due to the fact that there is a clearer separation between the migration of the cdmaOne air interface and that of its core network, making it possible for operators to upgrade their air interface technology and their core network almost independently of each other. cdmaOne's core network technology is based on yet another standard known as IS-41, which also happens to be the same core network standard used by TDMA. IS-41 already supports some limited packet-switching functionality. A number of operators, however, are in the process of upgrading their network and integrating simple Internet Protocol functionality, before eventually migrating to a full-blown version of the Mobile IP protocol. This full-blown version of Mobile IP, which we discuss in Chapter 4, "The Mobile Internet," is in many ways more advanced than that of GPRS—the GPRS Tunneling Protocol we alluded to earlier. This explains why some people refer to CDMA2000 1X as a 3G standard rather than a 2.5G one. As far as data rates are concerned, IS-95b can, in theory, support peak rates of 115.2 kbps—actual IS-95b services advertise peak rates of 64 kbps. In October 2000, Korean operator SK Telecom was the first to offer CDMA2000 1X services, quickly followed by LG Telecom and KT Freetel. As of late 2001, a number of other operators are in the process of deploying similar services that are expected to support peak rates of around 300 kbps—probably closer to 150 kbps in practice. The CDMA2000 1X air interface operates in the same 1.25-MHz bands as cdmaOne and is backward compatible with this standard, providing for a rather inexpensive upgrade.

The next migration step for most CDMA operators will bring them to full-fledged 3G through deployment of CDMA 2000 1X EV (EV stands for *evolution*). Some might, however, opt for WCDMA technology, which will likely gain broader acceptance than CDMA2000—thanks to harmonization efforts between the 3GPP and 3GPP2 partnerships, it is possible to run a WCDMA air interface on top of an IS-41 core network with full Mobile IP. 1X EV is essentially an overlay technology that will operate on top of the CDMA2000 1X infrastructure. In contrast to 1X, 1X EV can allocate users three 1.25 MHz channels and support peak data rates of up to 2.4 Mbps. Further enhancements are under discussion to support even faster data rates.

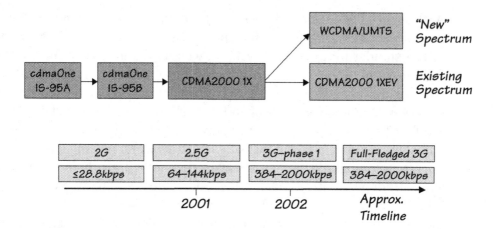

Figure 3.14 Possible CDMA migration paths.

Concluding Remarks

The number of possible migration paths to 3G can at times be overwhelming. Perhaps now is a good time to take a step back, look at the bigger picture, and see what it tells us when it comes to building m-commerce applications and services.

The good news is that, the overall situation is not quite as confused as it might seem at first glance. In fact, while the 2G landscape is a highly fragmented one with a number of competing standards—we only reviewed the main ones, but there are more such as PHS in Japan or iDEN in the United States—the overall 3G landscape is significantly simpler. There are essentially three emerging standards: two relying on CDMA technology—WCDMA/UMTS and CDMA2000—and one relying on TMDA—EDGE. Upon closer inspection, while there are three different air interfaces—four if we add the indoor TDD version of WCDMA/UMTS—there are only two core 3G network standards:

- An evolved GSM/GPRS core network standard
- An evolved IS-41 core network standard

Thanks to harmonization efforts between the 3GPP and 3GPP2 partnerships in charge of developing 3G standards, one can expect all three radio interfaces to eventually become interoperable with the two core network standards. More importantly, all 3G standards are compatible with the Internet Protocol (IP) and hence, at least in the long run, will make it much easier for people to develop mobile Internet applications and services.

Now, if this is the case, you might ask, "why all the fuss about all the possible migration paths?" As it turns out, the bad news is that the transition to 3G will not happen overnight. In fact, deployment of 2.5G and 3G technologies has been marred by delays. For example, in Europe, early GPRS deployment efforts suffered from an insufficient supply of handsets and excessively high pricing schemes by mobile operators. In Japan, DoCoMo's WCDMA deployment, which was first due in May 2001, had to be postponed until October of the same year due to numerous technical glitches. To make matters worse, while GPRS deployment is a financial no-brainer for most GSM operators, upgrading to UMTS/WCDMA is a rather expensive proposition. As a result, most operators are spreading their deployment plans over a number of years, first focusing on highly populated areas, where they are most likely to quickly recoup their investments. Thus, while the overall 3G landscape might look rather simple, the fact of the matter is that, for the foreseeable future, 3G will have to coexist with 2.5G and even 2G technologies. Even the UMTS forum, which includes some of the most ardent proponents of 3G, estimates that, by 2010, only 28 percent of mobile users will be connecting through 3G services—still, by then, this 28 percent should amount to about 600 million users. The consequence of all this for mobile application developers is that they will not only have to continue paying attention to the bandwidth requirements of their solutions, but will also have to worry about what happens when their users roam into 2.5G and even 2G areas.

By now, it should also be clear that there is no one preferred way of migrating to 2.5G or 3G, but a number of possible paths. In deciding which path to follow, operators need to consider a number of factors:

- **Legacy 2G technology.** The cost of each migration path is dependent on the legacy system you start with. When it comes to 2.5G, GSM operators are all migrating to GPRS, while cdmaOne operators are generally finding it much cheaper to migrate to CDMA2000.

- **Available spectrum.** While some upgrades can be done using existing spectrum, deployment of WCDMA/UMTS generally requires the acquisition of new spectrum licenses in the 2GHz band (see Table 3.1). In some countries such as Japan or Finland, this spectrum was made available to operators at no cost, provided they agreed to meet a number of conditions set by these countries' regulation authorities. Elsewhere, as in the UK or Germany, government auctions led license prices to skyrocket, leaving many operators strapped for cash and others without the required spectrum to upgrade their networks. In the United States, the situation is even worse and, at the time of writing, there are no clear plans to free spectrum for WCDMA/UMTS.

- **Roaming prospects.** In general, opting for one standard over another will limit an operator's ability to enter roaming agreements with others.

Table 3.1 WCDMA/UMTS license costs in different countries. The wide disparities in cost-per-capita show that migrating to WCDMA/UMTS is proving much more expensive in some countries than in others.

COUNTRY	ISSUE DATE	3G LICENSE COSTS	
		TOTAL COST ($BILLION)	COST PER CAPITA ($)
Finland	3/99	0	0
Spain	3/00	0.5	11.2
Britain	4/00	35.4	594.2
Japan	6/00	0	0
Netherlands	7/00	2.5	158.9
Germany	8/00	46.1	566.9
Italy	10/00	10	174.2
Austria	11/00	0.7	86
Norway	11/00	0.9	20.5
S. Korea	12/00	3.3	69.6
Australia	3/01	1.2	30.3
Singapore	4/01	0.2	42.6

- **Available resources.** At the end of the day, each operator has to see how much money it has available or is willing to spend on 2.5G and 3G upgrades. The path that might appear the most promising in the long run might prove to be infeasible, given available financial resources.

Nevertheless, one thing is clear: operators have aggressively started deployment of 2.5G technologies and, while deployment of 3G will be spread over many years to come, 2.5G already offers the prospect of always-on, packet-switched services and data rates several times faster than today's 2G systems. This will likely prove sufficient for many m-commerce applications and services to take off. After all, i-Mode has managed to secure 30 million users on 9.6-kbps packet-switched services.

The Mobile Internet

Introduction

M-commerce is about the provision of mobile Internet services. Unfortunately, the mobile Internet is not quite like its fixed counterpart.

For one, the variety of 2G, 2.5G, and 3G bearer services we reviewed in Chapter 3, "Mobile Communications: The Transition to 3G," requires the development of solutions that isolate application developers from the idiosyncrasies of these standards. As it so happens, the basic Internet protocol suite (TCP/IP), which was in part designed to play a similar role on the fixed Internet, was not developed with wireless links and mobile devices in mind.

The mobile Internet is in part about reconciling the many bearer services developed by mobile infrastructure providers, and the Internet protocols and Web standards produced by the Internet Engineering Task Force (IETF) and the World Wide Web Consortium (W3C). While these standards have started to converge—with, for example, the development of mobile IP and its integration in 3G standards or the move toward the next version of IP (IPv6)—this process will take many more years to complete. In the meantime, transition standards such as the Wireless Application Protocol (WAP) have been developed to bridge the gap. We will look at key characteristics of WAP, including its security features and how it compares with other approaches. As we will see, despite much criticism, WAP has emerged as a de facto standard of the early mobile Internet. It is, nevertheless, an evolving technology and we will discuss where it will likely fit in the future. We also look at the evolution of mobile application architectures as we transition from 2G to 3G and try to answer the question of how open the mobile Internet really is. This includes an overview of ongoing work by 3GPP on Open Service Access (OSA) and its impact on business models to come. We conclude with a review of 3GPP's Mobile Execution Environment (MExE), which provides ways of combining WAP with simple versions of Java.

As you read this chapter, keep in mind that there is more to the mobile Internet than reconciling different standards or providing for end-to-end security. The mobile Internet is also about recognizing the limitations of mobile devices and the new usage scenarios they entail. It is about rethinking the way in which we develop Internet applications and services to accommodate the input/output limitations of these devices, the time-critical nature of many of the tasks in which mobile users engage, and their limited attention as they talk to friends or drive through busy intersections. Developing services that accommodate these constraints requires moving away from wired Internet solutions such as search engines that return hundreds of hits and Web pages cluttered with banner ads. We will look at WAP usability issues and discuss the findings of a controversial study conducted by the Nielsen Norman Group in late 2000. We also introduce the concept of WAP User Agent Profiles, which provide a mechanism for tailoring presentation to the specificity of a particular device. Other usability issues are addressed in Part Three and, in particular, Chapter 7, "Next-Generation M-Commerce," where we discuss personalized, location-sensitive, and context-aware services.

Putting Things in Perspective

So far, we have been fairly informal in our treatment of mobile communication networks. In practice, however, because of the number of components and protocols that enter into play when building a network, people like to distinguish between different layers of functionality, with higher layers building on the functionality provided by lower ones. The most commonly used model for defining layers in a network architecture is known as the Open Systems Interconnect (OSI) model introduced in the mid 1970s by the International Standards Organization (ISO). In Figure 4.1, we can see a simplified version of this model, which will suffice for our discussion of the mobile Internet. The OSI model involves five layers:

Physical layer. This lowest layer is responsible for the conversion of bit streams into radio signals, and vice versa. While we alluded to some of the functionality of this layer in Chapter 3, a detailed discussion of it is not relevant to m-commerce.

Data Link layer. This layer deals with issues relating to access and sharing of the air interface, and includes, in particular, the different multiplexing schemes and standards we discussed in Chapter 3. Essentially, this is the layer responsible for establishing a connection between two nodes (or network devices) that are said to be on the same *link*; namely, that can talk to one another without any intermediary. An example is a mobile device and a base station.

Network layer. The Network layer builds on the Data Link functionality, to route packets through a network. In other words, it establishes a connection between two nodes, such as a mobile device and an Internet server, over multiple intermediate nodes (for example, routers). While this was not made explicit in our discussion in Chapter 3, we already saw that different mobile communication standards provide for very different levels of support here. We will now look more closely at these issues and review their implications.

Transport layer. This layer is responsible for making sure that packet transmission supported by the Network layer is reliable. This includes requesting that packets be re-sent, when they are lost or corrupted. The Transport layer is also in charge of flow and congestion control and related quality of service (QoS) issues.

Application layer. The Application layer, which actually encompasses several finer layers in the OSI reference model, is where people develop applications and services that take advantage of the communication functionality supported by lower layers. These range from simple functionality such as accessing one or more files (for example, HyperText Transfer Protocol (HTTP) and File Transfer Protocol (FTP)), to more complex ones such as videoconferencing, mobile payment, or location-sensitive services.

End systems such as the mobile phone and the Internet server represented in Figure 4.1 rely on protocols in all layers of the model. Intermediary systems, such as the transceiver base station in Figure 4.1, need only to worry about lower layers, as they do not implement any application but rather deal with infrastructure-level issues—in this example, routing traffic between the wireless link and the core communication network.

What Happens When We Try to Run TCP/IP on a Mobile Network?

The mobile communication standards we reviewed in Chapter 3 each have their own particular combination of protocols. At the Physical and Data Link level, they each rely on their particular way of modulating data or multiplexing data streams from multiple users. 2G standards such as GSM rely on circuit-switched connections, while 2.5G and 3G standards all rely on packet-switched networks. However, even among these latter standards, there are significant differences in the way in which packets are routed across the network. A major reason behind these differences has to do with the difficulty in adapting the standard Internet Transport

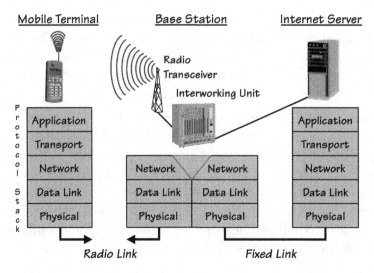

Figure 4.1 A simplified reference model.

and Network layer protocols (TCP/IP) to the demands of mobile communication networks. At the level of the Network layer, a new variation of the Internet's IP protocol is emerging, known as Mobile IP. Mobile IP is an integral part of each 3G standard, as already discussed in Chapter 3. However, packet-switched standards such as GPRS or CDMA2000 1X only support simplified variations of Mobile IP, and 2G protocols such as GSM have nothing equivalent. As far as the Transport layer is concerned, we will see that the standard TCP protocol used on the wired Internet leads to a number of problems, and that it is generally best to treat the wireless link differently from the rest of the network. We will look more closely at these and related issues. in the next section.

Mobile IP: Routing Packets to Mobile Nodes

Packets sent over the Internet contain headers, which among other things include their destination IP addresses. The Internet Protocol implements a decentralized approach to forwarding packets. Upon receiving a packet, a router looks up its destination address in a local routing table that tells it where the packet should be forwarded in order to eventually reach its final destination. For efficiency purposes, IP addresses are organized in such a way that nodes located in the same area of the network have identical prefixes. Thanks to this convention, routers can operate more efficiently by maintaining and looking up much more compact routing tables. This works fine as long as networks nodes remain static. As soon as a node, say a mobile phone or a PDA, starts moving around and attaching through parts of the networks with different prefixes, things start breaking down—especially if there are many mobile nodes such as the hundreds of millions of Internet-enabled mobile devices expected in the years ahead. This is because regular IP only offers two options for dealing with this situation, and neither is particularly satisfactory:

- One option is to start updating all routing tables with node-specific entries. Clearly, this approach does not scale up. Just think of the effort involved in constantly updating and searching tables with hundreds of millions of entries.

- The other option is to assign new IP addresses to the mobile node as it reattaches through different parts of the network. By doing so, we could at least ensure that the node's new IP address is compatible with the prefix of the network area through which it is currently connected. The trouble, however, has to do with packets that might already have been sent to the node's old address. What do we do with those? To make matters worse, other protocols such as TCP do not work very well when you start changing IP addresses like this.

To remedy these problems, a new variation of IP has been developed, called Mobile IP. The nice thing about Mobile IP—and also a key requirement when people started working on its design—is that it does not really require any hardware or software changes to the existing, installed base of IP version 4 (IPv4) hosts and routers other than those nodes directly involved in the provision of mobility services. Without going into the details, this is done using a process called *tunneling* and the concept of *care-of-address* (see Figure 4.2). Essentially, each mobile node keeps its IP address and relies on a *home agent* to intercept all packets sent to this address. As it reattaches to different parts of the network, the mobile node informs its home agent of its new location by sending it a care-of-address. The home agent encapsulates packets destined for the mobile node in new packets, which are forwarded to the care-of-address. This process is referred to as tunneling.

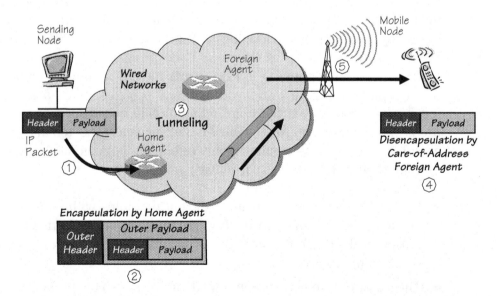

Figure 4.2 Overview of Mobile IP. A packet sent to a mobile node is intercepted by its home agent (1). The packet is encapsulated into an outer packet (2) for tunneling to the mobile node's current care-of-address (3). There, a foreign agent retrieves the original packet (4) and forwards it to the mobile node (5).

Because, to the network, encapsulated packets look like regular packets, they are forwarded to their care-of-address just as any other packet would be. Upon arrival at the care-of-address, a *foreign agent* (for example, a router) retrieves the original packet and forwards it locally to the mobile node. Clearly, all this needs to be done very carefully. For example, authentication procedures are required to make sure a malicious node does not attempt to pose as the mobile node and send its home agent a misleading care-of-address. As indicated earlier, Mobile IP is an integral part of all three 3G standards. This is not the case of intermediate standards such as GPRS or CDMA2000 1X, which implement limited variations of the protocol. Some elements of Mobile IP will also become available for free once IPv6 is more broadly adopted, although it is unclear how quickly this will happen.

TCP: The Wireless Link Requires Special Treatment

With Mobile IP, we can now rest assured that packets will be routed to the proper destination, even when that destination is a mobile device. IP, however, is a *best effort* service; it does not guarantee that packets will actually be delivered. This is one of the reasons why we need a Transport layer running TCP. TCP worries about making sure that all the packets are received, reassembled in the right order, and delivered to the proper application—say, HTTP versus FTP. This includes detecting lost packets and ensuring that they are re-sent. Another important aspect of TCP has to do with flow and congestion control. Specifically, TCP relies on the use of windows to control the number of packets (actually called *segments* in the TCP jargon) a particular sender is allowed to transmit without receiving an acknowledgment from a receiver. These windows have played a critical role in ensuring that overall Internet performance degrades gracefully, and that all users equally share in the pain that might result from a congested network. This is done by dynamically controlling the size of the window. In general, when things work well and acknowledgments come back, TCP will increase the size of the window allowing the sender to transmit at a faster rate. If, for some reason, acknowledgments fail to come back in time—based on roundtrip time estimates—TCP will assume that the network has become congested, or possibly that the receiver's buffer is full, and that it is time to show some good behavior. TCP then reduces the sender's window size. It does so in a fairly drastic manner, bringing the window size down to a minimum—a process known as *slow start*. Over time, if things work well and acknowledgments start coming back in time, the window size will be increased again. This approach works well in wired networks where the main cause for missing acknowledgments is congestion. Over wireless links, however, lost or corrupted packets are more common and congestion is rarely the problem. Instead, connections are frequently dropped, transmission errors are more common, and, to make matters worse, roundtrip times are harder to predict due to greater variations in latency. Often, when acknowledgments fail to come back, the correct reaction is not to reduce the transmission rate; rather, it should be to keep it the same or perhaps even increase it to make up for lost time. Imagine, for example, that you just wanted to download the latest stock news on your mobile phone as your car entered a tunnel—perhaps not the safest thing to do while driving. As you emerge from the tunnel and your connection gets reestablished, the last thing you want is the server on the other side to slow the rate at which it sends your stock update. Yet, by default, this is what TCP would do. A number of different solutions have been proposed to remedy this situation. They all involve differentiating between the wireless link and the rest of the network in one form or another. One solution, known as Indirect TCP, simply partitions the connection and relies on the foreign agent, introduced in the context of Mobile IP or some equivalent node, to act as an intermediary, sending its own acknowledgments (see Figure 4.3). This allows for transmission parameters such as the window size to be set differently over the wired network and the wireless link, although it introduces other complications. For example, when the mobile node moves and reattaches to a different foreign agent, all the packets buffered by the old foreign agent need to be forwarded to the new one, a possibly lengthy process (see Figure 4.4). Other solutions such as the one imple-

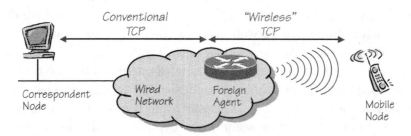

Figure 4.3 One way to adapt TCP to a mobile environment is to effectively break the connection into two, one over the wired network and the other over the wireless link, each with its own TCP parameters. This solution is known as Indirect TCP.

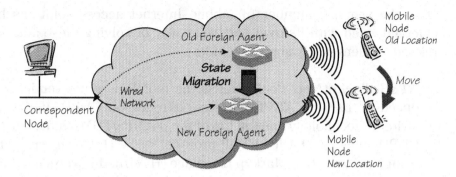

Figure 4.4 Indirect TCP can entail long state migrations when a mobile node reattaches to a different part of the network.

mented in GPRS rely on retransmission protocols at the level of the Data Link layer to quickly re-send lost packets over the wireless link.

Another potential problem with TCP is its relatively complex setup procedure, which involves getting the sender and receiver to agree on different parameters such as the proposed maximum packet size or the initial transmission window. In doing so, seven packets need to be exchanged. On the wired Internet where latencies are on the order of a fraction of a second, this is hardly a problem. Over a wireless link, where latencies are often measured in seconds, sending packets back and forth can result in intolerable delays.

The Mobile Internet: Early Precursors

There is much more to the mobile Internet than adapting TCP/IP to the specificity of mobile communication networks. Other challenges involve developing presentation languages that accommodate the limited screen size, memory, and power of mobile devices, and the limitations of mobile networks. Even simple protocols such as HTTP, which is normally used to handle the transfer of Web pages over the wired Internet, is not particularly well suited to mobile environments, where the loss of connections is a rather common occurrence. This is because HTTP is a stateless protocol. It has no memory, and hence cannot pick up from where it was in the event you lose your connection.

Over the years, a number of mobile Internet access solutions have emerged, each with its own particular way of solving these and other related challenges. Examples include:

- **AT&T PocketNet.** PocketNet is the mobile Internet service introduced in 1996 by AT&T Wireless. It relies on CDPD, the packet-switched technology built on top of the D-AMPS/TDMA infrastructure. CDPD includes a basic TCP/IP protocol stack. However, rather than using the HyperText Markup Language (HTML), the Web's most common presentation language, and a traditional Web browser, PocketNet relies on a stripped-down markup language known as the Handheld Device Markup Language (HDML) and a specialized microbrowser, UP.View. HDML and UP.View were developed by a company called Unwired Planet—later renamed Phone.com and eventually acquired by OpenWave. Besides limiting the type of content that can be displayed—mainly just text—HDML was the first language to introduce the *deck of cards* metaphor. This metaphor stems from three simple observations:

  - Mobile devices have tiny screens, and hence can only display so much content at any one time.

  - Many user applications can be broken down into small sets of interactions (*cards*) with the user, each involving a fairly small number of options (for example, selecting from a menu with three or four options, or typing a user login and then a password).

  - By bundling together each of these sets of interactions (into a *deck of cards*) and sending all associated screens at once to the mobile device, it is possible to significantly reduce the number of times the user will have to wait for data to be transferred over the mobile network—often a rather frustrating experience over 2G networks.

  - HDML eventually evolved into the Wireless Markup Language (WML), the presentation language of the Wireless Application Protocol (WAP).

- **Palm.Net—WebClipping.** Another early mobile solution was Palm.Net, a Web service introduced in 1998 by Palm for its PDAs. Palm.Net was initially launched on top of BellSouth's Mobitex data network and uses a solution known as WebClipping. WebClipping optimizes data transfer of HTML documents by allowing static Web content to be stored on the PDA while only having dynamic information sent over the wireless link. Several hundred Web services have been adapted to run on Palm.Net, ranging from enterprise applications to finance, shopping, information, and entertainment services (see Figure 4.5).

■ **NTT DoCoMo i-Mode.** With about 30 million users at the end of 2001, i-Mode is by far the most popular mobile Internet service today. At the time of writing, the system still operates on top of DoCoMo's PDC-P packet-switched network (also known as DoPa), which has a built-in TCP/IP stack. For content presentation, it relies on Compact HTML (cHTML), essentially a stripped-down version of HTML 3.2 designed by Japanese company Access Corporation. Users who connect to i-Mode have access to official content provider sites through DoCoMo's i-Mode servers that take care of authentication and billing on their behalf (see Figure 4.6). They can also access nonofficial content providers—essentially, any Internet site with content written in cHTML, although subject to significantly weaker security guarantees. While i-Mode has often been pitted as a competitor to WAP, the reality is that NTT DoCoMo is a very active member of the WAP Forum,

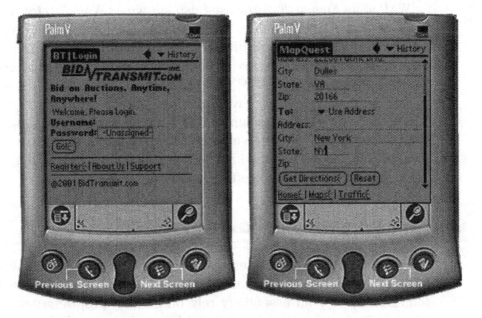

Figure 4.5 Examples of WebClipping services.

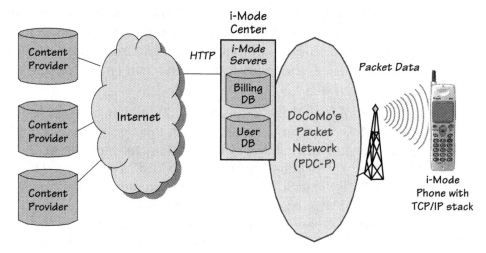

Figure 4.6   Overall i-Mode architecture.

eager to leverage its mobile Internet experience in markets outside of Japan by moving toward the global WAP standard. This is best exemplified by i-Mode's recent move from cHTML to XHTML Basic. XHTML Basic is the new content presentation standard adopted by WAP in its August 2001 release (also known as WAP 2.0) for replacement of its earlier Wireless Markup Language (WML). Other recent i-Mode enhancements include the introduction of faster PDC-P data rates of 28.8 kbps. The service is also expected to be introduced over DoCoMo's nascent WCDMA network. Finally, in early 2001, i-Mode also started supporting Java-based applications through its *i-appli* service, which relies on Sun's Java2 Micro Edition (J2ME), making it possible to provide end-to-end security and support more dynamic or more interactive applications such as games, zoomable maps, services providing regular traffic updates, and so forth.

Enter the Wireless Application Protocol

While each of the mobile Internet access services we just reviewed has had its fair share of success, as exemplified by i-Mode's impressive customer base, they all essentially rely on ad-hoc, proprietary solutions often limited to specific devices and bearer services. In contrast, the

WAP Forum, founded in 1997 by Phone.com, Ericsson, Nokia, and Motorola, aimed from the start at developing an open standard that would be both device- and bearer-independent. Over the years, the Forum has grown to more than 600 members. Together, the WAP forum members represent 99 percent of the global mobile handset market. The Forum also includes most of the leading mobile operators, including Japan's NTT DoCoMo. The commitment of its members to adopt WAP in their products and services has made it a de facto standard. By offering a standard that operates across most bearer services, from SMS to 3G, and across a broad range of devices, WAP makes it possible for content providers and other mobile Internet players to achieve badly needed economies of scale.

Despite these advantages, ever since its first release in 1998, WAP has been the subject of criticism. Advocates of Internet orthodoxy have complained that it was too far a deviation from W3C and IETF standards. Developers found its content presentation features to be too constraining and its approach to security questionable. Users and people in the media have blamed it for a poor overall user experience. The reality is that, while WAP is far from perfect, a number of its limitations have been addressed in more recent releases of the standard. Today's latest release, WAP 2.0, provides support for end-to-end security, color graphics, animation, push technology (making it possible, for example, to send news alerts and location-sensitive ads to mobile users), and a number of other features that were not initially present. The WAP Forum has also learned to collaborate more closely with other standardization bodies such as W3C and IETF. This has led to the adoption of XHTML Basic (also known as WML2) as its new presentation language, and to the support of the TCP and HTTP protocols. In summary, with a number of 2G, 2.5G, and 3G bearer services likely to coexist for many years to come, WAP looks like it is here to stay for now. This is not to say that it will not evolve or that it should not be supplemented with other technologies such as Java.

Overview of the WAP Architecture

WAP is not a protocol, but rather a suite of protocols aimed at bridging the gap between the variety of mobile bearer services and basic Internet protocols such as TCP/IP and HTTP, the HyperText Transfer Protocol commonly used to access Web pages over the fixed Internet.

WAP Gateway and Legacy Protocol Stack

Until the release of WAP 2.0 in the summer of 2001, WAP always required the introduction of a WAP gateway or proxy. The gateway's role includes interfacing between the WAP protocol stack, which is specifically designed for operation over the wireless link, and the regular Internet protocol stack (see Figure 4.7).

Because of bandwidth limitations, WAP content is encoded into a compact binary format before being transmitted over the wireless link. The WAP gateway is responsible for decoding it into text that can be interpreted by HTTP, and vice versa. In other words, when a user attempts to access a particular Web site from his mobile device by entering its URL address, his WAP device first encrypts the request into WAP's compact binary format and then sends it to the WAP gateway over the wireless link. The gateway converts it into a regular HTTP request, which is for-

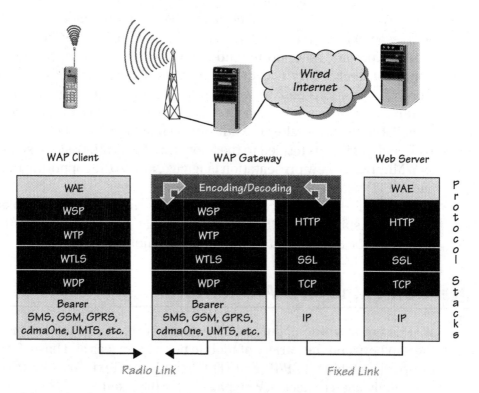

Figure 4.7 The WAP gateway with WAP's legacy protocol stack.

warded to the URL's server over the fixed Internet. The server then returns a WML/WML Script (or XHTML Basic) document to the gateway, which encrypts it and forwards it to the mobile device. Do not worry for now about all the acronyms in the WAP protocol stack—we will look at them in a moment.

More generally, the WAP gateway helps improve communication efficiency by caching content between the server and the mobile device. It is also used to authenticate users, making sure they have a plan that includes access to WAP services, and often provides support for additional billing functionality.

Operating without a WAP Gateway

With the release of WAP 2.0, the original suite of wireless protocols is now referred to as the WAP *legacy stack*. This is because, as a result of close cooperation with IETF, WAP 2.0 now supports an alternative stack, which is available for fast bearer services that already include the IP protocol—mainly 2.5G and 3G bearer services. This alternative stack essentially consists of variations of the TCP and HTTP protocols whose parameters have been specifically configured (*profiled*) for operation over a wireless link. The wireless version of HTTP also provides for compression of content and the establishment of secure tunnels between the mobile device and the Web server. When using this protocol stack, it is, in theory, possible to do away with the WAP gateway. In many cases, however, the gateway is still needed, as it helps optimize communication and can also provide support for additional functionality such as billing, location-based services, or privacy features. A WAP gateway is also required for push services.

A Closer Look at the WAP Protocol Stacks

The legacy WAP protocol stack, which was the only one available before the release of WAP 2.0, has been optimized for low-bandwidth network services with relatively long latencies. It is the only option available for bearer services that do not support an IP stack. For fast bearer services with built-in IP such as GPRS or WCDMA/UMTS, WAP 2.0 offers an alternative stack of protocols that implements versions of the TCP and HTTP protocols optimized for operation over the wireless link. Let's now take a closer look at the protocols in each stack (see Figure 4.8).

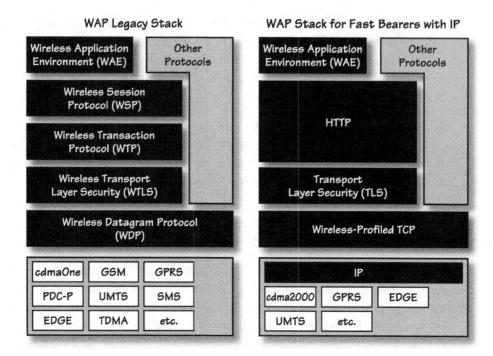

Figure 4.8 The two WAP protocol stacks.

WAP Legacy Protocol Stack

The legacy stack comprises the following protocols:

Wireless Datagram Protocol (WDP). This lower-level protocol serves as an interface between the underlying bearer service and the upper layers of the WAP protocol stack. Due to the variety of bearers supported by WAP (from SMS and GSM all the way to WCDMA/ UMTS), the functionality embedded in this protocol can vary greatly. In the case of bearers with built-in IP such as GPRS, WDP simply implements the User Datagram Protocol (UDP), a bare-bones connectionless protocol that does not worry about sequencing or loss of packets. UDP simply ensures that packets are properly delivered to the upper layers. Retransmission of lost packets is handled at a higher level by the Wireless Transaction Protocol (WTP).

Wireless Transport Layer Security (WTLS). The WTLS layer is designed to provide privacy, data integrity, and authentication between two communicating applications. This layer is essentially based on the

Transport Layer Security (TLS) protocol commonly used over the fixed Internet—TLS was formerly known as the Secure Socket Layer (SSL) protocol. However, while on the fixed Internet, TLS operates on top of TCP; WTLS has to take care of some of this functionality on its own, making it different from TLS. In addition, WTLS has been designed to operate over low-bandwidth, high-latency wireless links. This includes support for variable levels of security, each involving different hand-shake requirements—with higher levels of security requiring more complex handshake procedures and hence more bandwidth. The different encryption mechanisms supported by WTLS also vary in terms of key length, making it possible to also adjust to the computing power available on the mobile device—longer keys require more demanding computations. Security is far from a trivial issue over low data rate bearer services such as GSM or cdmaOne (not to mention SMS) and mobile devices with limited memory and processing power. Developers have to be very selective in deciding how much security is really required by their applications. To make matters worse, because WTLS is different from TLS, its fixed Internet counterpart, it is impossible to provide true end-to-end security with the WAP legacy stack. Instead, processing on the WAP gateway includes decryption and re-encryption of information between the two protocols. As a result, for a very short moment, information on the WAP gateway is vulnerable to interception, making the gateway a potential security hole in the WAP legacy protocol (see Figure 4.9). For this and other reasons, ownership of the WAP gateway is an important factor in determining the overall level of security of WAP applications. Most applications today rely on gateways

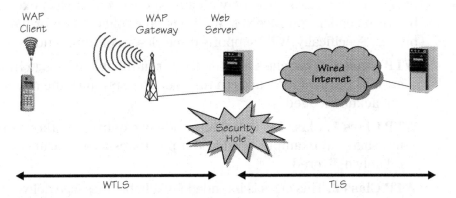

Figure 4.9 When using the WAP legacy stack, the gateway represents a potential security hole.

operated by mobile operators, wireless ASPs, and other trusted third parties such as banks. As we will see, the gateway security hole can be eliminated when using the alternative protocol stack introduced in WAP2.0. However, this option is only available for fast bearer services with built-in IP.

The Wireless Identity Module

WAP includes an option for enhancing WTLS security features with the use of a tamper-resistant device referred to as the Wireless/WAP Identity Module (WIM). This module, which is similar in philosophy to the SIM module introduced in the GSM standard (see Chapter 3), is most likely to be implemented using a smart-card that could be inserted in the mobile device. This could actually be the same smart card used for the SIM module in GSM. Remember that all 500+ million GSM phones today already have such a SIM module. GPRS and UMTS also have a SIM, and standards such as CDMA are in the process of introducing equivalent modules, as they provide for a particularly secure and versatile way of storing user information and cryptographic keys. These security issues are discussed in more detail in Chapter 5, "Mobile Security and Payment."

Wireless Transaction Protocol (WTP). WTP has been designed as a light-weight transaction-oriented protocol for implementation in *thin clients*; namely, mobile devices with little memory and processing power. It essentially acts as a substitute for TCP, while also handling some HTTP functionality, and has been optimized for operation over the wireless link. This includes handling acknowledgment and retransmission of lost packets. However, in contrast to TCP, which involves a relatively complex setup process with a total of seven packets exchanged between sender and receiver, WTP has no explicit setup or teardown phases. Specifically, WTP supports three classes of transaction services:

WTP Class 0. This class is intended for unreliable message transfer, such as pushing messages to the user. Reception of the message is not acknowledged.

WTP Class 1. Class 1 is for reliable message delivery without a result message. An example would be a push message that needs to be reliably delivered.

WTP Class 2. This class is intended for reliable message delivery with a result message confirming reception. It is particularly useful for banking applications, as it allows, for example, transaction rollbacks.

Wireless Session Protocol (WSP). WSP provides HTTP/1.1 functionality. While HTTP is stateless, WSP introduces the notion of a shared state between a client and a server to optimize content transfer. The process by which clients and servers agree upon a common level of protocol functionality during session establishment is referred to as *capability negotiation.*

The shared client-server state in WSP can also be used to resume suspended sessions in exactly the same context in which they were suspended. When browsing, this means that you can resume your session exactly where you left it when you switched off your phone or lost your connection. WSP is also capable of managing multiple sessions, allowing users to suspend one and switch to another. For example, you could switch from a GPRS session used to read your email to a UMTS session for a videoconference with an important client, and later return to reading and writing email exactly where you left off.

Finally, WSP is also responsible for the binary encoding of sessions, and supports both pull and push functionality.

Wireless Application Environment (WAE). WAE is the most visible part of the WAP protocol stack to both users and developers. It includes the WAP microbrowser, the WAP markup and scripting languages. Another important part of WAE is the Wireless Telephony Applications Interface (WTAI), which allows users to access telephony applications from WAP sessions. Because of the importance of this layer, it is the subject of a separate subsection later in the chapter.

WAP Protocol Stack for Fast Bearers with Built-In IP

This new stack is similar to the wired Internet protocol stack, except that it relies on parameter settings that have been optimized for communication over the wireless interface. Specifically, it comprises the following protocols:

- **Wireless Profiled TCP (WP-TCP)** is a version of TCP optimized for wireless environments based on recent recommendations from the IETF. It is fully interoperable with its fixed Internet counterpart.

- **Wireless Profiled TLS** is a wireless profile of the TLS protocol found on the wired Internet. In contrast to WTLS, which has to make up for some missing TCP functionality in the legacy stack, the Wireless Pro-

User Agent Profiles

Another interesting feature introduced in WAP 2.0 is referred to as User Agent Profiles, namely a protocol to specify the capabilities of the WAP client (or *user agents*) that builds on the Composite Capability/Preference Profile (CC/PP) standard defined by W3C. User Agent Profiles can be used to represent information such as:

- Hardware characteristics of the mobile device (for example, screen size, color capabilities, image capabilities, support for Bluetooth, and so forth).
- Software characteristics (for example, operating system, list of audio and video encoders, support for Java, and so forth).
- Application/user preferences (for example, browser type, supported scripting languages, but also whether the user wants the sound of his device to be on or off).
- WAP characteristics (for example, WAP version, maximum WML deck size—see the subsection on WAE).
- Network characteristics (for example, bearer characteristics such as latency and reliability).

Using this information—which is typically transmitted in WSP headers (or HTTP headers when using the new stack)—the Web server can much more accurately determine how to best tailor content to the capabilities of the particular mobile device. User Agent Profiles are allowed to change during a session. For example, if the user roams into a new area with very different network characteristics, the Web server could decide to switch to a text-only session, removing all graphics in the content sent to the WAP client. Similarly, if the user places his handset into the phone cradle in his car, the server could start sending audio information instead of visual information.

filed version of TLS allows for TLS tunneling between the mobile device and the Web server. Since now there is no need to decrypt and encrypt information on the WAP gateway, this implementation eliminates the security hole associated with the WAP legacy stack, making it possible to support true end-to-end security (see Figure 4.10).

- **Wireless Profiled HTTP** is a fully interoperable version of HTTP/1.1 that supports message body compression and the establishment of secure tunnels.

- **Wireless Application Environment**: This layer is identical to the one implemented in the legacy stack, and is further detailed in a separate subsection later in the chapter.

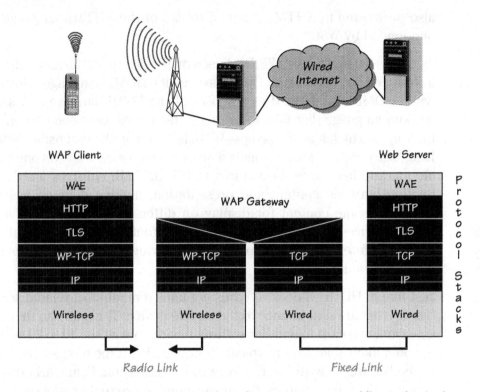

Figure 4.10 Example of TLS tunnel through a WAP gateway, enabling end-to-end security between a WAP client and a Web server.

The Wireless Application Environment

The Wireless Application Environment (WAE) is the part of WAP that users see. It consists of:

- The WAP markup and scripting languages
- The WAP microbrowser
- The Wireless Telephony Applications Interface

The WAP Markup and Scripting Languages

WAP content was originally required to be written using what is now referred to as the legacy Wireless Markup Language (WML)—sometimes also denoted WML1 or WML1.x—and the WML Scripting Language, WMLScript. With the release of WAP 2.0 in August 2001, content can now also be written in XHTML Basic, a subset of the XHTML language recommended by W3C.

As discussed previously, WML is based on the *deck of cards* metaphor first introduced by Phone.com/OpenWave in its HDML language. However, while HDML was defined as a subset of the HTML language—a sloppy markup language that fails to properly distinguish between content and the way in which it is displayed—WML is based on the extensible Markup Language (XML). XML is actually a *meta-language*, as it allows one to create markup languages. In contrast to HTML, XML enforces a strict distinction between content and presentation, making it much easier to adapt the same content for display on different devices (for example, mobile devices with different screen sizes, fonts, and color capabilities) or even different media (for example, displaying information versus reading it using voice synthesis).

Just like in HDML, a deck of cards contains a number of related screens (*cards*) that are sent at once to the mobile device. The user can then navigate from one card to another without having to wait the 10 to 20 seconds it would likely take over a typical 2G connection if the next screen had to be fetched from the Web server over the wireless link. Figure 4.11 displays a simple deck with two cards—never mind the actual syntax of the language; there are enough books on WML. In general, the developer can present the user with a number of options, allowing him to jump from one card to a (small) number of others. In this particular example, however, the user is only given one choice, which is to move to Card 2. This is done by first clicking on the left soft key on his mobile phone (this key is not displayed here, although it would be right below where *Options* appears on the first screen) and then selecting *Next* on the following screen, using the same soft key. Soft keys are simply keys whose function (*semantics*) can be dynamically altered by the program running on the device. Note that the code in this example is static. In general, just as dynamic pages can be generated by a regular Web server, a WAP server can create dynamic WML pages on-the-fly; for example, using the Web's Common Gateway Interface standard. This is particularly useful if you want to develop services that will query databases, using input provided by the user or based on user profile information. In fact, it is hard to think of useful services that would rely solely on static pages—what good would they be?

WAP also supports the addition of *client-side* logic to Web pages through its WMLScript language, which is itself essentially a bare-bones version of JavaScript. WMLScript allows developers to process informa-

WML *Code*   **Display**

```
<?xml version="1.0"?>
<!DOCTYPE wml PUBLIC "-//WAPFORUM//DTD WML 1.1//EN"
"http://www.wapforum.org/DTD/wml_1.1.xml">

<wml>
<card id="Card1" title="WML Primer">
    <do type="accept" label="Next">
      <go href="#Card2"/>
    </do>
    <p>
       Select Next to go to Card 2.
    </p>
  </card>
  <card id="Card2" title="WML Primer 2">
    <p>
       This is Card 2.
    </p>
  </card>
</wml>
```

Card 1

--WML Primer--
Select Next to go
to Card 2.

Options

-Browser Options-

Next

Select   Back

--WML Primer2--
This is Card 2.

Card 2

Options

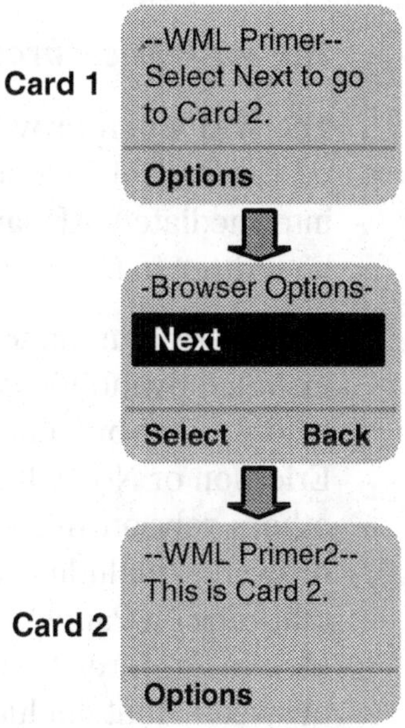

Figure 4.11 Example of a two-card WML deck. The options displayed in bold at the bottom of each screen are selected using the mobile device's soft keys.

tion entered by the user on the client side, rather than having to send this information all the way to the WAP server for processing. This is particularly useful when implementing forms, where you want to make sure that the information entered by a user is valid; for example, making sure locally that, when prompted for his age, the user actually enters a number. WMLScript comes with a library of functions. Of particular importance is the WMLScript SignText() function used to implement digital signatures in mobile devices. Its use is further discussed in the section *WAP Security* in Chapter 5.

Finally, with the release of WAP 2.0, WAP now also supports XHTML Basic. This is a stripped-down version of XHTML, the Extensible Hyper-Text Markup Language recommended by W3C as the language of choice to bridge the gap between the Web's easygoing days of HTML and the cleaner way of developing Web content provided by the Extensible Markup Language (XML) and Cascading Style Sheets (CSS)—CSS define the way in which content is rendered. By converging with the language recommended by W3C, WAP reduces training time and makes it possible for developers to write applications for both PCs and WAP clients—or at least re-use more code than in the past.

The WAP Microbrowser

The WAP microbrowser sits on the mobile client. Its main role is to interpret and display content received from the Web server—possibly via an intermediate WAP gateway. With the introduction of WAP 2.0, content can be written in the legacy WML language, WML Script, or XHTML Basic.

There are a number of WAP microbrowsers on the market, some preinstalled by mobile device manufacturers and others available for downloading on your mobile device. Several device manufacturers such as Ericsson or Nokia have developed their own WAP microbrowsers, while others rely on third-party software providers. Third-party WAP microbrowsers include Openwave's Mobile Browser, AU-System's browser that runs on PalmOS, Windows CE, and EPOC platforms, and Microsoft's Mobile Explorer. Some companies such as 4thpass are also offering WAP browsers that include stripped-down versions of Java. For example, 4thpass's KBroswer runs Sun Microsystems' Java Kilo Virtual Machine (KVM), the most compact version of J2ME available (see the section *A Word about Java and MExE*).

The Wireless Telephony Applications Interface

The Wireless Telephony Applications Interface (WTAI) contains a list of APIs that allow developers to invoke telephony applications from the WAP browser. Imagine, for example, that you just connected to a WAP city guide to get a list of nearby restaurants. Using WTAI, developers can return a list of restaurants, each with a click-through option allowing the user to call and reserve a table without having to quit the WAP browser. Using WTAI, the developer can also offer the user the option of saving the number in his phone book, and so forth. Some WTAI features are available across all mobile networks, while others might be network/operator specific.

The WAP Usability Saga

WAP introduces a number of features that are intended to make mobile Internet services more usable. They include the *Deck of Cards* metaphor in WML, the introduction of persistent sessions in WSP, the use of WAP User Agent Profiles to better tailor content, and the integration of WAP with telephony applications through the WTAI interface. Despite these efforts, early surveys of WAP users revealed a high level of frustration. Major sources of complaints have included:

- Slow connections, which also translate into steep bills when users get charged by the minute, as is the case over circuit-switched networks such as GSM.
- Deadends or sites being down.
- Sites with poor signposting, making it difficult for users to get the information they are looking for.
- Uneven content quality such as news sites displaying outdated headlines.
- Small screens.

One of the most rigorous studies of early WAP users was conducted by Web usability guru Jakob Nielsen and associate Marc Ramsay of the Nielsen Norman Group. The study, which was carried out in late 2000 in the London area, involved a group of twenty users who were asked to access WAP services over GSM over a period of one week and record their impressions in a diary. Traditional usability tests were also performed at the beginning and end of the study, where users were timed as they carried out different tasks. The study showed that mundane tasks such as retrieving news headlines, checking the local weather forecast or accessing TV program listings were taking significantly longer than most users could tolerate. Worse, the times failed to significantly improve even after a week of daily use, as users continued to confront slow connections and poorly designed sites. Overall, 70 percent of the study participants indicated that they did not see themselves use WAP within the next year. However, this figure dropped to 20 percent when they were asked to think over a three year horizon, suggesting that, despite their frustrations, they generally saw potential in the mobile Internet.

By and large, the problems uncovered by this and other similar studies of early WAP users fall into four categories:

1. Technical limitations, which were more often related to the underlying GSM network than to WAP itself.

2. Poor site design, which often made inefficient use of the limited screen real estate and failed to provide intuitive navigational support to users (for example, pages spanning multiple screens with no indication that the user should scroll). Site design generally reflected wired Internet practices with little or no attempt to accommodate the constraints and usage scenarios associated with a new medium—some sites looked like they had simply translated their HTML content into WML.

3. Poor content and poorly managed sites: Content should be compelling enough to make up for the frustrations associated with mobile access devices and slow data rates. In contrast to i-Mode, many European GSM operators initially failed to set up business models and quality control standards that would help generate the necessary critical mass of quality content.

4. Device limitations: While they can be expected to become less severe over time with the introduction of more powerful devices and larger screens, these limitations will continue to make the mobile Internet a very different medium from its wired counterpart, requiring the use of particularly careful design methodologies.

The Nielsen survey created quite a stir when it was first published. By and large, it indicated that, as long WAP users would connect over GSM networks, they were unlikely to be very satisfied. The good news is that none of the problems uncovered in the study are insurmountable. Far from that, the introduction of GPRS and other 2.5G and 3G packet-switched technologies can be expected to alleviate the main source of user frustration by providing for faster, always-on connections and more affordable packet-based billing schemes. Business models similar to i-Mode's can help build a critical mass of more compelling content. Careful design practices also need to be adopted that recognize the specificity of the mobile Internet and the wide variety of usage scenarios it entails. This includes understanding the broader context within which the user operates as well as carefully analyzing the steps required to accomplish different tasks. The conclusions of the Nielsen survey are actually quite promising. They suggest that a large percentage of users is likely to adopt the mobile Internet when the problems identified are remedied. This finding is supported by the i-Mode experience and its 30 million users. When all the right ingredients are in place, users will flock to the mobile Internet, as they have in Japan. Just as we do, Ramsay and

Nielsen identify two particularly promising categories of mobile applications and services:

1. Killing time—currently by far the most popular set of services on i-Mode's portal.

2. Highly personalized, context-aware services capable of supporting users as they engage in time-critical, goal-driven tasks—the subject of Chapter 7.

The Nielsen Norman study, as many others, also raises some issues about the design of WAP. Designers of the WAP legacy protocol stack made a number of sacrifices to accommodate slow mobile communication standards such as GSM and even SMS—think about WTLS for instance. If WAP ends up being unusable over GSM, what was the point of making all these sacrifices in the first place? The good news is that, to a great extent, this question is becoming moot with the introduction of the new WAP2.0 protocol stack.

Evolving Application Architectures—
How Open is the Mobile Internet?

The fixed Internet is characterized by an open architecture that allows just about anyone to set up shop online. This open architecture has been key to the rapid growth of the Internet, leading to the emergence of a *click-through* economy where value chains can be modified and new partnerships entered into by seamlessly redirecting online customers from one web site to another. This level of openness is very different from the one found on the emerging mobile Internet today, although efforts such as those under 3GPP's Open Service Access (OSA) initiative are currently under way to change this situation—this does not mean that additional standardization efforts are not also required over the fixed Internet.

Despite the introduction of standards such as WAP, today's mobile Internet is still in many ways constrained by the rigid architectures of 2G and 2.5G communication networks. Under these legacy architectures, critical network functionality and user information are buried in the operator's backbone network—for example, see the GSM Home Location Register in Figure 3.9. Access to the user's location, SMS functionality, billing support services and authentication functionality, to name just a few, are all under the tight control of the mobile operator and are not standardized. The end result is a high barrier to entry for third parties who want to develop applications and services capable of running across multiple mobile communication networks. While this situation contributes to the strategic place mobile operators occupy today in the m-commerce value chain, it is far from ideal. Operators need third-party developers and content providers to develop new applications and services if they are to increase their data revenues.

This situation is set to change over time with the emergence of architectures such as OSA—Open Service Access, formerly known as *Open Service Architecture* in the UMTS specifications. The OSA architecture is developed by 3GPP in collaboration with Parlay, a multi-vendor consortium eager to promote the development of open, network-independent APIs to access critical network functionality, and a number of members of the JAIN Community, a group of companies working on the development of Java-based APIs for the telecom industry.

OSA is a component-based architecture for future *service networks*, where applications and services from the operator and third-party developers can be flexibly assembled through standardized interfaces. A key element of the OSA vision is the notion of *Virtual Home Environment* (VHE), namely the idea that independently of which network operator they connect through, roaming users should have access to the same set of services and these services should have the same look and feel.

The OSA service network, which is based on IP, builds on a set of service enablers expected to grow over time, and a Personal Service Environment that helps personalize applications and services:

1. **Service Enablers:** These components add standardized functionality to the basic bearer services provided by a mobile network.

   ■ **Service Capability Servers** give applications standardized access to features that the core network provides. A mobile positioning server is a typical example of a service capability server. It makes it possible to build location-sensitive applications and services (for example, an E911 service) that can access a user's location without having to worry about which network the user is connecting through and which particular positioning technology it relies on. Examples of such solutions will be described in Chapter 7. Another typical example of a Service Capability Server is a WAP gateway that makes it possible, for example, to push content to users, independently of the underlying bearer service. In OSA, WAP is actually part of a broader service capability known as the Mobile Execution Environment (MexE), which is discussed later in this chapter.

- **Application Support Servers** are service enablers that do not directly rely on features built into the core network. Instead, they provide access to additional components such as billing gateways, email servers, or public key infrastructure components. A typical example is a charging support server that outside applications and services such as games or traffic report services, can access through a standard API to post charges to the user's bill. This makes it possible to support a variety of billing schemes from charging per transaction, per click, per game, and so forth. The charging support server essentially performs billing mediation, collecting detailed usage statistics and passing them to the billing gateway, which in turn converts them into standardized Charging Data Records (CDRs) that can be interpreted by the operator's legacy billing system—originally built with voice traffic in mind.

2. **Personal Service Environment (PSE).** This is a set of databases that store subscriber information and preferences, which can be selectively accessed by different applications and services through standardized APIs. The PSE helps create a Virtual Home Environment, where users are presented with the same set of personalized applications and services independently of the network through which they connected. A more detailed discussion of the PSE and related solutions such as Microsoft's Passport is provided in Chapter 7.

The resulting OSA model is one where applications and services, as well as elements of the service network itself, can easily be developed by third parties through standardized APIs. This presents mobile operators with a number of possible business models to choose from (for example, whether to focus solely on voice and data transport functionality, or also provide elements of the service network such as authentication, positioning and personalization functionality). It also makes it possible for new entrants to position themselves as value added intermediaries by offering a layer of service enablers and personalization functionality without owning any wireless infrastructure. Virtual Mobile Network Operators, mobile portals, third party location information brokers, or third party billing and payment providers are different variations of this model, as discussed previously in Chapter 2, "A First Look at the Broader M-Commerce Value Chain."

A Word about Java and MExE

As discussed earlier, WAP derives part of its appeal from its device and bearer independence. Device independence (or *Write Once, Run Everywhere*) has also been one of the objectives of Java, although for a long time this looked more like a distant dream for devices such as mobile phones. With the advent of more powerful mobile devices, what once seemed like a dream is slowly emerging as an increasingly viable option. In 1999, Sun Microsystems introduced a version of Java specifically designed to run on mobile devices, Java 2 Micro Edition (J2ME). With J2ME it becomes possible to offer significantly higher levels of interactivity at the level of the mobile client—without having to communicate back with the server (for example, mobile games, interactive maps, simple agents that regularly provide you with updates such as stock market or weather updates, and so forth). By implementing TLS, Java can also provide support for end-to-end security between the client and server.

J2ME currently comes in two flavors or configurations:

- A Connected Device configuration (CDC)
- A Connected Limited Device configuration (CLDC)

Each configuration corresponds to a minimum set of device requirements in terms of memory and microprocessor characteristics. CLDC is the more minimalistic of the two, which makes it particularly suitable for mobile phones and mainstream PDAs. It relies on a particularly small virtual machine known as the K Virtual Machine (KVM), where K is used to indicate that the machine's footprint is measured in kilobytes rather than megabytes—somewhere between 40 and 80 kilobytes, depending on different compilation options and the device on which the virtual machine is installed. A special set of device-specific Java APIs known as the Mobile Information Device Profile (MIDP) has also been developed for CLDC, providing for a complete J2ME runtime environment particularly well-suited for mobile devices. The first Java-enabled mobile phones started becoming available in Korea in late 2000. In early 2001, i-Mode launched its *i-appli* service featuring over a hundred different Java applications. By late 2001, over 3 million Java-enabled mobile phones had already been sold, and most device manufacturers were offering or had announced plans to offer J2ME-enabled handsets.

The resulting situation is one where both WAP and Java are emerging as particularly attractive application development environments for future mobile devices. Nevertheless, with mobile devices sporting an ever broader range of options, the 3GPP consortium felt that this might not be sufficient to avoid a fragmentation of the application market. To prevent this from happening, they have developed specifications for 3G Mobile Execution Environments (MExE) that define three broad categories of devices or *classmarks*. The expectation is that developers will create applications that can run on any device within one or more of these classmarks, making it much easier for someone to determine whether a given application is suitable for a particular device.

- Classmark 1 corresponds to devices that only support WAP1.1 or later versions (which are all backward compatible).
- Classmark 2 requires devices that run Personal Java, a rather demanding version of the language for a mobile device, as it uses over 1 megabyte of memory.
- Classmark 3 corresponds to an intermediate category of devices capable of running kJava, the CLDC version of J2ME with the MIPD profile.

Concluding Remarks

The mobile Internet is about the convergence between the many 2G, 2.5G, and 3G communication standards and Internet protocols that were not originally designed for mobile, wireless environments. WAP has been developed as an open standard that aims to bridge this gap and isolate developers from the idiosyncrasies of a highly fragmented mobile communication landscape. While WAP has been the subject of much criticism, the latest version of the protocol has gone a long way to address many of the problems identified when it was first introduced.

Thanks to close cooperation between the WAP Forum, the IETF, and W3C, Internet protocols are being adapted to the demands of mobile communication. WAP 2.0 has also introduced a new protocol stack that relies on versions of the HTTP/TCP/TLS/IP specifically configured to accommodate the communication over the air interface. This stack, however, is only available for fast bearer services with a built-in IP. As a result, in many cases, issues of end-to-end security will continue to require special attention, at least in the near future. Beyond WAP, Java is also expected to play an important role in the development of mobile applications and services, and several million Java-enabled devices are already in use today. Finally, there is more to the mobile Internet than WAP and Java. The mobile Internet is also about the creation of an environment where key network functionality such as user location, authentication, or billing can be made accessible to third party developers through standardized APIs. Efforts such as 3GPP's Open Service Access (OSA), which are working on the specification of such APIs, are expected to play a key role in opening the mobile value chain to third-party players and in creating an explosion in the number of m-commerce applications and services.

CHAPTER 5

Mobile Security and Payment

Introduction

Without trust and security, there is no m-commerce—period. How could a content provider or payment provider hope to attract customers if it cannot give them a sense of security as they connect to paying services or make purchases from their mobile devices? Consumers need to feel comfortable that they will not be charged for services they have not used, that their payment details will not find their way into the wrong hands, and that there are adequate mechanisms in place to help resolve possible disputes. As these examples show, trust and security have many dimensions—some technical, others related to our perception of how safe a given environment is or how convenient a given solution appears to be. Secure solutions such as Public Key Infrastructure (PKI) have been slow to gain acceptance among the broad public because they are perceived as hard to understand and somewhat more difficult to use. While cryptography is central to security both on the fixed and mobile Internet, its complexity is generally well hidden from the eyes of the user—when it is not, its acceptance is proving significantly more difficult.

As it turns out, beyond the added risk of forgetting your handset in a taxi, the mobile Internet introduces a number of additional challenges over its fixed counterpart. These challenges are mainly related to the limitations of mobile devices and the nature of the air interface over which transmission takes place. Even typing your name, credit card number and its expiration date—by far, the most common form of payment over the fixed Internet— is not a viable option when considering the input capabilities of most mobile devices. When using cryptography, the limited memory and processing power of most handsets severely restrict the types of algorithms and the lengths of keys that can be used. As we saw in earlier chapters, unless special precaution is taken, communication over the air interface is more vulnerable to eavesdropping, and the low data rates and frequent disconnects of the mobile Internet have led to standards such as WAP that do not necessarily guarantee end-to-end security. To make matters worse, as we receive emails, download songs and pictures, and run Java applets on our mobile devices, they are no longer immune to viruses and worms. In June 2000, VBS/Timifonica was the first worm script reported to target mobile phones, sending *I Love You* style messages from PCs to GSM handsets. With an increasing number of companies purchasing mobile devices to enable their workforce to connect remotely to their intranets, mobile security is no longer just a consumer market concern but a high corporate priority. The good news is that, when carefully deployed, existing mobile technologies can generally provide for adequate security.

Over the past couple of years, a number of operators have discovered that the authentication mechanisms and billing infrastructure they have set up puts them in an ideal position to serve as micropayment providers and collect fees on behalf of content providers. A central element of this infrastructure in standards such as GSM, GPRS, or UMTS is the Subscriber Identity Module (SIM), which is under the control of the operator and is used to store authentication keys. As already discussed in Chapters 3 and 4, this module, which is also finding its way into CDMA, is typically implemented through a smart card inserted in the handset. WAP has also introduced its own version of the SIM in the form of a Wireless Identity Module or WIM, which could either be implemented on the same card as the SIM or on a separate card—one that would not necessarily be issued by the operator but perhaps by a bank, credit card company, or some other third-party player.

In general, ownership and control over the method used to authenticate the user is at the core of a number of competing payment solutions. They include, among others, solutions involving the use of a second card that are generically referred to as *dual slot handset* solutions. In the second part of this chapter, we will review several standardization initiatives in this area, including the Mobile Electronic Transaction (MeT) forum, the Mobey initiative, the Global Mobile Commerce Interoperability Group (GMCIG), the Mobile Payment Forum, and Radicchio, and see how they relate to each other. As we will see, the ambition of some of these efforts goes beyond the support of wireless (and fixed) online purchases. It includes the development of true digital wallets that will combine all our credit cards, debit cards, and other methods of payment to provide for *point-of-sale* (POS) payment solutions, possibly using technologies such as Bluetooth. Finally, we will see that a number of payment alternatives developed for the fixed Internet are also being adapted for use over mobile communication networks.

As we review different aspects of mobile security, it is important to keep in mind that security always requires an overall approach. At the end of the day, a system is only as secure as its weakest component, and securing networking transmission is only one part of the equation. The sad truth is that people often prove to be the weakest link in the chain, whether in the form of a rogue employee who hacks the company's billing database or WAP gateway, or a careless user who writes his PIN number on the back of his GSM handset and forgets it in the subway.

Revisiting Security: The Role of Cryptography

In earlier chapters, we touched on a number of different security issues. We discussed the role of the SIM module and Authentication Centers in the GSM architecture in Chapter 3, "Mobile Communications: The Transition to 3G"—a similar solution is implemented in GSM evolution standards such as GPRS. We reviewed mobility issues in the context of IP and TCP and their counterparts in the WAP legacy stack. We noted the possible security hole at the level of the WAP gateway and discussed how the new WAP protocol stack attempts to remedy this situation for fast bearers with built-in IP. In this section, we take a closer look at the role of cryptography in supporting these and related aspects of security, and some of their business implications.

Technically speaking, there are a number of different dimensions to network security, each corresponding to a different class of threat or vulnerability. Protecting against one is no guarantee that you will not be vulnerable to another. When looking at network transmission, one often distinguishes between the following essential security requirements:

- **Authentication** is concerned about verifying the identities of parties in a communication and making sure that they are who they claim to be.

- **Confidentiality** is about ensuring that only the sender and intended recipient of a message can read its content.

- **Integrity** is concerned about the content and making sure that what was sent is also what is received—the content of your messages and transactions should not be altered, whether accidentally or maliciously.

- **Non-repudiation** is about providing mechanisms to guarantee that a party to a transaction cannot falsely claim later that she did not participate in that transaction.

Cryptography plays a central role in satisfying these requirements. There are, however, other mechanisms that contribute as well (for example, packet acknowledgments or checksums).

Cryptography is essentially about taking ordinary *plaintext* messages and turning them into scrambled ones or *cipher text* using various algorithms. The recipient can, in turn, decipher the scrambled message and recover the original plaintext message using a matching decryption algorithm. In order to keep things secret, both encryption and decryption algorithms rely on one or more keys. Keys are generally kept secret (or at least one key in the case of Public Key Cryptography) and are combined with the messages that need to be encrypted or decrypted. Therefore, even if the encryption and decryption algorithms are made public—which is generally the case in open Internet standards—no one can read their content unless they have access to the secret key. Encryption algorithms rely on mathematical properties to also ensure that trying to guess the key or the content of the encrypted message is nearly impossible or at least would require so much time that no one would want to bother.

Secret Key Cryptography

Secret key cryptography, also known as *symmetric cryptography*, relies on the same key for both encryption and decryption (see Figure

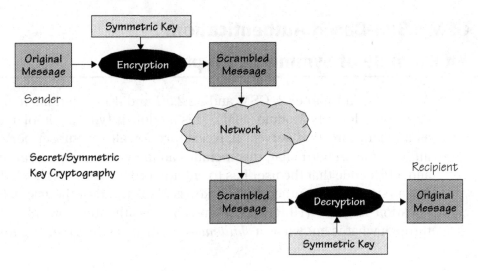

Figure 5.1 Private/symmetric key encryption.

5.1). It works well if people can securely agree on a secret key ahead of time. However, when you try to communicate over the Internet with people you had never thought of before, you might be faced with a chicken-and-egg problem. How do you securely agree on a key with them if you do not already have a secure means of communication? Storing ahead of time different keys for every person you might one day want to communicate with does not sound very practical, and using the same key with more than one person raises non-repudiation issues. If several people use the same key to communicate with you, how can you tell which one actually sent you a message? As we will see in a minute, public key algorithms provide for a much more straightforward way of dealing with key exchange—a particularly important issue over the Internet, where you constantly interact with new people. Secret key cryptography has the merit, however, of being much faster than public key cryptography. For this reason, public key algorithms are often used to exchange secret keys at the beginning of secure sessions, with data transmission during the session actually relying on symmetric cryptography, using the key that was just exchanged.

GSM's SIM-Based Authentication: An Example of Symmetric Cryptography

As we saw in Chapter 3, GSM and its 2.5G and 3G evolutions include a Subscriber Identity Module (SIM). This module is typically implemented as a smart card that serves as repository for all the subscriber's vital information. In particular, this includes an individual authentication key and a PIN code that the user has to enter in order to unlock the SIM. The same (symmetric) authentication key is also stored in the user's Home Location Registry (HLR) (see Figure 3.9). Authentication takes place through what is known as a *challenge-response* method (see Figure 5.2).

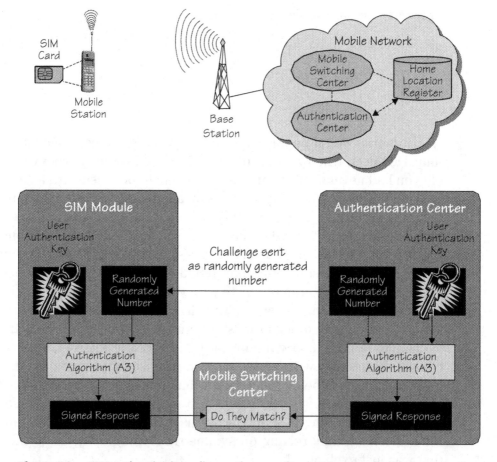

Figure 5.2 GSM authentication relies on the use of a secret authentication key stored on both the user's SIM card and in the user's HLR on the network.

Essentially, the authentication center sends a random number to the SIM over the wireless link. The SIM combines this number with the user's authentication key using an authentication algorithm known as A3. This computation is performed on the SIM itself, so the key never leaves the SIM (in other words, it never ends up on the mobile device, but stays on the smart card inserted in the device). The result of this computation, called a *signed response*, is returned over the wireless link for comparison with the result obtained by the authentication center, carrying out the exact same operation with the copy of the key maintained in the HLR. If the two values match, the user is considered authenticated. Actual communication takes place through symmetric encryption, using a specific *session key*, which both the authentication center and the SIM generate based on another random number, known as A8. This particular session key is passed by the SIM to the mobile station (see Figure 5.3). This way, encryption during the session can be more efficiently carried out on the device itself. The session key itself is used in combination with yet another (symmetric) encryption algorithm, referred to as A5 in the GSM standard. It is also shorter than the authentication key to further speed up processing during the session itself.

An interesting feature of this approach is that it solves the secret key exchange problem by requiring users to have a SIM module, which they obtain when they open their mobile phone subscription with the mobile operator. This situation puts the mobile operator in a position of strength when it comes to negotiating partnership arrangements with third-party content providers; although, as we will see shortly, alternative authentication solutions that do not necessarily require going through the mobile operator are starting to emerge.

Public Key Cryptography

In *public key cryptography*, encryption and decryption rely on related, yet different keys: a public key, which is available for anyone to see, and a private key, which the user will keep to herself. The interesting property of this pair of keys is that to decrypt messages encrypted with one, you need the other. The keys are said to be *asymmetric*. For example, if I want to send you a message, I can get your public key to encrypt the content of my message and, using your private key (the one that is only known to you), you will be the only one capable of decrypting it (see Figure 5.4).

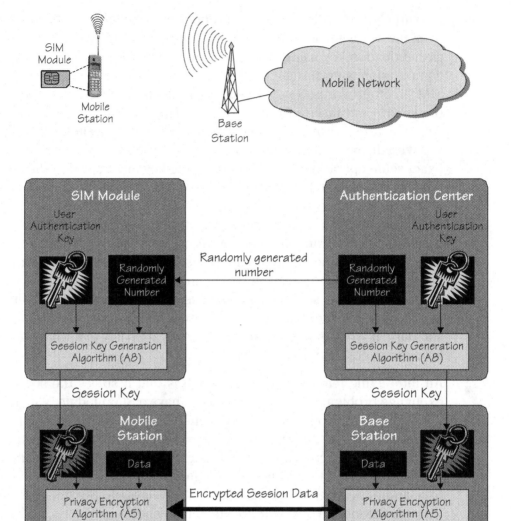

Figure 5.3 Generation of a GSM session key.

The most popular algorithm for public key cryptography is RSA, named after its inventors, Rivest, Shamir, and Adleman. It works with variable key sizes—longer keys require more memory and more processing, but provide for stronger security. More recently, *elliptic curve cryptography* algorithms are starting to find their way into mobile devices. They rely

A Word about Smart Cards

Smart cards are microcomputers that are small enough to fit in a wallet or even a mobile phone. They have their own processors and memory for storage. Communication with a smart card is defined through an interface that specifies what can be accessed and how. As a result, smart cards can be made very secure—they are generally characterized as *tamper-proof*, and are ideal to hold cryptographic keys and run cryptographic algorithms. While a number of smart cards rely on proprietary operating systems, a small number of operating systems are starting to emerge that aim at handling multiple applications on the same smart card. They include Mutlos, developed by smart card manufacturer Mondex; and Microsoft Smart Cards, designed to only work with other Microsoft operating systems. Javacard, which relies on a Javacard Virtual Machine that can run on top of different smart card operating systems, offers a "write once, run anywhere" solution to smart card application development. The GSM standard and its SIM module have also led to the specification of a SIM Tool Kit (STK) for SIM application development, which supports communication between SIM applications and the keypad and display of GSM/GPRS phones.

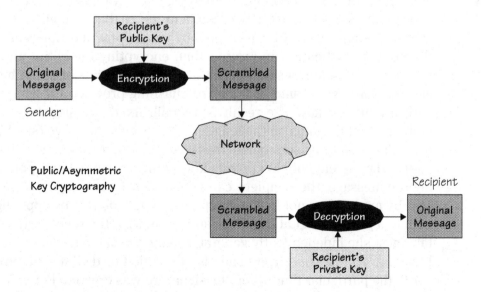

Figure 5.4   Public/asymmetric key encryption.

on different mathematical properties that allow for shorter keys—a particularly appealing feature on mobile devices. As always, acceptance of new cryptographic algorithms takes time. Even when they appear much better on paper, people are often concerned that they have not been subject to the many years of intense scrutiny older algorithms have had to survive through actual use.

Digital Signatures

Note that in the scenario just described, there is no guarantee that the message you will decrypt is the exact same one I sent you, assuming that I even sent you one. Someone might have altered my message, or it might actually be coming from a third party that just pretended to be me. If you are a mobile operator with whom I have an account and someone else orders 10 movie tickets over his mobile phone pretending to be me, I will not be very happy when I get my phone bill at the end of the month. To prevent this from happening, we need *digital signatures*. Specifically, suppose that before encrypting my message with your public key, I first encrypt it with my own private key. The message resulting from this first encryption step—the one with my private key—is equivalent to a signed message. While anyone can decrypt it, using my public key, only I could have created it, since I am the only one who knows my private key. Therefore, by combining both keys, my private key first and then your public key, I can actually create a message that can only be attributed to me and can only be decrypted by you. In practice, things are a little more sophisticated, and rather than encrypting the entire message with my private key, a small digest—*message digest*—of it can be created (according to some standardized hashing process) for encryption with my private key. The result is a small, fixed-size signature rather than one that depends on the length of the original message (see Figure 5.5). The resulting scrambled message can only be opened by the intended recipient with her private key—only known to her. Upon opening the message, the recipient can check that the message was indeed sent by me and has not been corrupted. She can do this by applying my public key to the digital signature and comparing the result with a digital digest she independently generates using the same hashing function I used. A *digital timestamp* can also be added to digital signatures to verify the particular time when the signature was created. For example, if after ordering 10 movie tickets over my mobile phone and apposing a

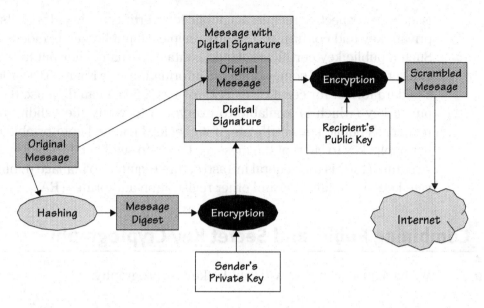

Figure 5.5 Adding digital signatures allows for authentication of the sender and for checking the integrity of the message by the recipient. Here we only illustrate the steps on the sender's side.

digital signature to the order I were to claim that someone had stolen my private key, a timestamp might help determine whether the transaction took place before or after the time I claim my private key was stolen. Timestamps also provide protection against someone intercepting my message and re-sending it a few days later, ordering another 10 movie tickets. By inspecting the timestamp on the message, my bank or the movie theater could determine that something is wrong. This is known as *replay protection*.

Certificate Authorities

This is all fine, but suppose that someone gains access to the directory where my recipient keeps her public key and substitutes it with his. He could now possibly intercept messages intended for her, alter them and re-encrypt the resulting messages with my recipient's actual public key. The recipient would not see the difference. This is why we also need *certificate authorities* (CAs) or trusted third parties (public or private), as they are also referred to. A well-known example is VeriSign. A CA can issue my prospective recipient a public key certificate signed with its own private key and confirming that my recipient's public key is indeed hers. Such a public key certificate would contain her name, her public key, an expiration date, and some additional information (see Figure 5.6)—the ISO standard format for certificates is known as X.509. I can then use the CA's public key (which is available to everyone) to verify the validity of my recipient's public key certificate. If it checks, I can feel comfortable using her public key to encrypt the message I want to send her. Public Key Infrastructure (PKI) is the general infrastructure required to manage public/private keys, including CAs and other registration authorities (RAs).

Combining Public and Secret Key Cryptography

We have just seen that with public key cryptography:

- Authentication of the sender of a message can be performed using digital signatures, which are obtained by encrypting message digests.
- Confidentiality is achieved by encrypting a message with the recipient's public key, which is itself certified by a CA.

Figure 5.6 A digital certificate is signed by a CA and can be verified using the CA's public key. The certificate typically includes several keys; for example, one for digital signatures and one for key exchange.

- Integrity of the message is guaranteed using digital signatures, since they are tied to the content of the message or at least a digest of it.

- Non-repudiation is also provided by digital signatures, typically in combination with timestamping.

Because public key cryptography is relatively slow, it is generally used only for authentication and the exchange of symmetric session keys. Information exchanged during the session itself is encoded with the symmetric key. The resulting combination also provides for authentication, confidentiality, integrity, and non-repudiation. This is the approach taken in the Internet's Transport Layer Security (TLS) protocol, which we introduced in our discussion of WAP 2.0 in Chapter 4, "The Mobile Internet."

Message Authentication Codes

In communication, checksums are often used to verify the integrity of messages. Checksums, which are obtained by applying simple algorithms to the content of a message, are sent along with the message itself. When the recipient receives the message, she can apply the same algorithm to the content of the message and check the result against the checksum that was sent along with it. If they match, there is a decent chance that the message was not corrupted. *Message Authentication Codes* (MACs) are based on the same principle, but use symmetric key cryptography to protect against both accidental and malicious message tampering. The symmetric key is exchanged at the beginning of a session, typically using public key cryptography, and is applied to message digests. The result is sent along with the message. By applying the same algorithm at the other end, using the same symmetric key, the recipient can verify both the authenticity and integrity of the message—which explains why MACs are also referred to as *Message Integrity Codes* (MICs).

The Combinations Are Many

There are many security protocols and cryptographic algorithms used over the Internet—whether fixed or mobile. A complete review of all the possible variations, even just within the context of WAP, is well beyond the scope of this book. These many variations reflect the need to selectively adapt security to the nature of the data being exchanged. In 1999, the IETF finalized specifications of the IP Security (IPSec) protocol, which since then has also been adopted in standards such as GPRS. The protocol provides for negotiation of security parameters between a sender and a recipient in the form of a *security association*. A security association can be thought of as a table with a number of entries, each corresponding to a different set of security parameters (for example, specific encryption algorithm, key, and so forth). Depending on the sensitivity of the data they exchange, the sender and receiver can select among these different entries to decide how to protect their communication. Negotiation is carried out using the Internet Key Exchange (IKE) protocol (also known for historical reasons under the acronym ISAKMP/Oakley), which allows for the use of digital certificates for device authentication.

Flexibility in adapting security parameters is particularly critical in mobile environments, given the limited memory and processing power available, especially when keys are stored and processed on smart card modules such as a SIM or a WIM. The low bandwidth and high latencies associated with wireless links are further constraints and call for minimizing the number of packets exchanged as part of the negotiation process itself.

Revisiting WAP Security and the Role of the WIM Module

While we already covered all the key elements of WAP security in Chapter 4, now is perhaps a good time to recap how they all fit together. In WAP, security is essentially provided through the combination of the WTLS/TLS protocols (depending on the particular WAP protocol stack in use), the WIM module, and WMLScript SignText() (see Figure 5.7):

- **WTLS/TLS.** When using the WAP legacy stack, the only one available in the absence of a fast bearer with built-in IP, security is provided through WTLS, a WAP-specific security protocol. WTLS, which in many ways resembles TLS, is however different, requiring decryption and re-encryption at the level of the WAP gateway. The resulting security hole creates a situation where ownership of the gateway is critical in determining the overall security of the system, as already

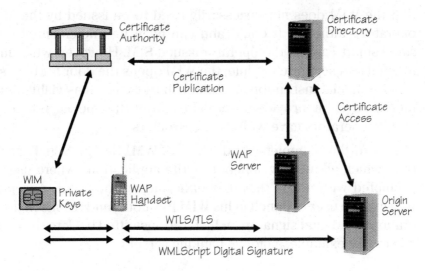

Figure 5.7 WAP security—the big picture.

pointed out in Chapter 4. When running over a fast bearer with built-in IP, WAP 2.0 allows you to use TLS over the wireless link, thereby eliminating this security problem. Both WTLS and the version of TLS supported by WAP 2.0 allow for several different security levels, and are generally configured to accommodate the limitations of the mobile device and the wireless interface. WTLS, for example, supports three security classes:

- Class 1 uses no certificates, and hence provides no authentication.
- Class 2 requires a server certificate, but no client certificate.
- Class 3 corresponds to the highest security level, requiring certificates from both the client and server.

WTLS also relies on optimized handshake procedures—which are in part required to make up for the absence of a TCP equivalent in the WAP legacy stack.

- The **Wireless Identity Module** (WIM) can be used to hold private and secret keys required by WTLS and TLS, as well as by non-WAP applications. It can also store certificates, although those do not need to reside on a tamper-resistant device such as the WIM and can in fact simply be stored in a directory elsewhere. The critical element here is that the WIM does not necessarily need to be issued by the mobile operator—although it could and can even be implemented on the same smart card as the operator-issued SIM. However, when implemented as a separate module, the WIM opens the door for banks and other financial institutions to establish themselves as viable m-commerce payment providers—circumventing the special relationship mobile operators have with their customers.

- This is further complemented with the **WMLScript SignText** function, which allows developers to write applications where users are prompted with a text that they either accept or reject. Acceptance requires the user to punch in his WIM PIN code, and results in the generation of a digital signature, which is transmitted back to the content server (for example, a mobile banking server).

Mobile Payment

Payment is one of those areas in which the mobile Internet is quite different from its fixed counterpart, both in terms of new constraints it imposes and new opportunities it offers:

Constraints. Today's most common method of payment for one-time purchases over the fixed Internet—entering your credit card details into your computer and transmitting this information over a secure TLS connection—is not the most practical option when using a mobile phone. The devices' tiny keypads do not provide for a particularly pleasant experience, and obtaining a secure end-to-end TLS connection is not always an option. In addition, many transactions over mobile networks involve rather small sums of money—*micropayments*. Think 50 cents for a horoscope, a dollar for a traffic update or a few dollars for a bus pass or movie tickets.

Opportunities. At the same time, as we develop secure mobile payment standards, our mobile phones are progressively turning into *personal trusted devices (PTDs)* that we may come to view as substitutes for our wallets and the many different credit cards and debit cards they so often contain. Using technologies such as Bluetooth, we may start employing them as point-of-sale methods of payment just as we would use cash or credit cards today.

The constraints we just identified have allowed mobile operators to take advantage of their existing billing infrastructure and their control over user authentication mechanisms such as the SIM module to position themselves as preferred mobile payment providers. Under this model, they take responsibility for billing the user on behalf of merchants and service providers and, in the process, often take a cut of every transaction. This is the i-Mode model we have so often referred to in this book (see Chapters 1 and 2)—although i-Mode does not rely on SIM modules for user authentication. The mobile operator's position of strength is viewed with envy by many other players and is being challenged in the context of new mobile payment solutions. One of the most commonly proposed solutions involves using the WIM card as a second identification module that would be issued by a bank or some other third party. This option is commonly referred to as a *dual-chip* solution.

Figure 5.8 outlines the many different alternatives identified by the Mobey Forum to implement an additional security element in mobile handsets—Mobey is a grouping of financial institutions and handset manufacturers eager to influence the development of mobile payment standards. Approaches other than dual-chip solutions involve the use of a second slot with a card reader, which can either be integrated in the handset (dual-slot handset) or be a separate device connected to the handset (for example, via Bluetooth). In 1999, Motorola introduced a STAR Tac Dual Slot GSM handset capable of reading third-party smart cards. The handset and its smart card reader have been used in two different ways in France and the UK:

- In the UK, mobile operator BTCellNet has offered the phone as a method for users to load electronic cash on smart cards, which they can then use to pay for purchases at participating stores. Solutions like this one, which involve reloading cash on a smart card, are commonly referred to as *electronic purses* or *e-purses*.

- In France, where credit cards such as Visa have been required for many years to be equipped with a smart card under a standard known as *Carte Bancaire* (which simply means "Bank Card" in French), a different approach has been offered by France Telecom's Itineris in its Itiachat pilot. Here, the mobile handset is used as a mobile point-of-sale terminal. Specifically, users indicate their intention to use this method of payment for purchases at participating Web sites by communicating their mobile phone number. In return, they receive an SMS

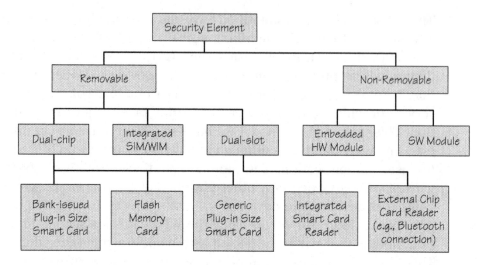

Figure 5.8 Alternative implementations of a mobile device's security element considered by the Mobey Forum.

message containing the details of the transaction, which they can either accept or reject. Acceptance is indicated by inserting the card into the second slot and entering a PIN code, which in turn results in an authorization request sent to the card issuer in the form of another SMS message. Eventually, a confirmation that the transaction has been completed is returned. More recently, this solution has also been demonstrated using the new *EMV* standard developed by Europay, MasterCard, and Visa to add smart cards to their credit cards.

Dual handsets, however, are not widely available (except in France) and tend to be relatively bulky, which might be a major impediment to their broader adoption. Dual-chip handsets appear to have a better chance of success—whether based on a WIM module or on inserting some other smart card or chip in the handset.

As of late 2001, the Electronic Payment Systems Observatory (ePSO) had already identified over 30 different mobile payment solutions, each with its own particular set of technologies (see Table 5.1 for a subset). Some solutions are provided by mobile operators, some by financial players, and yet others involve alliances between operators and financial organizations. Fixed-Internet payment solutions are also finding their way into the mobile world as examplified by Paypal or PayDirect! Virtual POS solutions are limited to online purchases from mobile phones, whereas real POS solutions refer to the local purchase of goods or services such as settling a restaurant bill, paying for parking, or buying an item from a vending machine. Some solutions also allow for peer-to-peer payments between individual customers. Independently of the particular technology they use for communication (for example, WAP, SMS, GSM's Unstructured Supplementary Services Data channels, or simply a voice channel), most solutions involve a relatively similar process:

1. The customer gets redirected by the merchant to the payment provider's server. In some cases, the customer directly contacts the payment provider, or the merchant tells the payment provider to contact the customer.

2. A more or less secure connection is established, whether via SMS, USSD, WAP, or over a voice channel, and the user is prompted to confirm the transaction details.

3. Payment authorization is typically entered in the form of a PIN code.

4. The customer and merchant later receive confirmation of the transaction.

Solutions provided by mobile operators tend to focus on micropayments in the context of both virtual and real POS. Operators have risk management procedures that are adequate for dealing with small amounts of money, whether through pre-paid or post-paid accounts. When it comes to dealing with larger amounts, they would, however, need to adopt tighter procedures for assessing, controlling, and monitoring risk (both at the level of customers and merchants), and might, in some countries, be required to acquire a bank license. Accordingly, the more ambitious initiatives that aim at handling both small and large payments tend to involve partnerships between operators and financial organizations.

Despite the large number of mobile payment solutions that have emerged over the past couple of years, mobile payment is a market where there is little room for fragmentation. Consumers will want solutions that they can use independently of the network they connect to, especially in places like Europe, home to nearly 60 different mobile operators. Mobile payment scenarios involving roaming users will also require that mobile operators generalize their current voice roaming

Table 5.1 Sample of Mobile Payment Solutions as of Late 2001

MOBILE PAYMENT SOLUTION	GEOGRAPHICAL COVERAGE (AS OF LATE 2001)	DESCRIPTION	ADOPTION LEVEL	PROVIDER	COMMENTS
Banko.max	Austria	Virtual POS		Max.mobil (owned by German T.Mobile)	• WAP mall • Digital wallet with links to credit cards and bank accounts • Launched in late 2000
Bibit	Netherlands, Belgium, Germany, Scandinavia, and so forth	Virtual POS		Bibit in partnership with KPN Mobile	• KPN WAP mall • Digital wallet aggregates a number of payment methods • Multi-currency payment
Cingular DirectBill	USA	• Virtual POS • Limited to micropayments (under $10) that are included on the user's phone bill		Cingular	• A separate credit card-based ewallet is offered for larger purchases in partnership with Bell South and SBC
EMPS	Finland	• Virtual POS with plans for a future real POS solution using Bluetooth or infrared	Pilot with 150 dual-chip phones (SIM+WIM)	Nokia, Visa, Nordea	• Pilot launched 9/01 • WAP, EMV, SET, Bluetooth, Visa Open Platform • Uses an EMV WIM card and combines credit, debit, and loyalty cards • Close links with Mobey Forum

Continues

Table 5.1 *(Continued)*

MOBILE PAYMENT SOLUTION	GEOGRAPHICAL COVERAGE (AS OF LATE 2001)	DESCRIPTION	ADOPTION LEVEL	PROVIDER	COMMENTS
GiSMo	Sweden, UK, Germany	Virtual POS		Millicom International Cellular	• Launched 11/99 • GSM/SMS based • Monthly electronic bill presentment allows users to pay using credit cards or bank transfers
iCash	Sweden	Real and virtual POS(limited to micropayments)			• GSM/SMS
i-Mode	Japan	Virtual and real POS (monthly billing)	30 million	DoCoMo	• Launched 2/99
Metax	Denmark	Real POS (gas stations)		Metax	• To be launched in 2002 • GSM based • Replaces Metax credit card payment
Mint	Sweden	Real and Virtual POS			• Parking service launched 1/01 • Virtual POS to be launched Q1/02
MobilPay	Germany and Austria	Virtual POS		Mobilpay.com	• Launched 6/00 • PIN returned over SMS

Continues

Table 5.1 Sample of Mobile Payment Solutions as of Late 2001 *(Continued)*

MOBILE PAYMENT SOLUTION	GEOGRAPHICAL COVERAGE (AS OF LATE 2001)	DESCRIPTION	ADOPTION LEVEL	PROVIDER	COMMENTS
Mobipay (aka MovilPago)	Spain	• Real (for example, taxis) and virtual POS • Both micro–payments and larger payments		**Operators:** Telefónica Móviles, Airtel, Amena **Banks:** BBVA, BSCH, Caja Madrid, Banco Popular, Banesto, Banc Sabadell	• Pilot: Q4/01, launch Q1/02 • Solution provider: ACI Worldwide • Uses SIM • PIN sent over GSM's USSD control channels • Merged with Pagomovil
Orange Mobile Payment	Denmark	Virtual POS		Orange	• WAP and SIM
Paiement CB sur Mobile (ItiAchat)	France	Virtual POS and mail order	120,000 dual-slot handsets sold by late 2000	Initially offered by France Telecom (Itineris). All other French operators are expected to eventually offer it.	• Dual-slot handset • SMS and now also WAP • Carte Bancaire or EMV card • Pilot in late 1999 • Launched in 2000
PayBox	Germany, Austria, Spain, Sweden, UK	Real and virtual POS and P2P	500,000 registered users and thousands of merchants	PayBox.net AG **Shareholders:** Deutsche Bank, Debitel, Mobilkom Austria	• Pilot 12/99 • Launched 5/00 • Fees vary per country (for example, merchants pay 3% of transaction in Germany)

Continues

Table 5.1 (Continued)

MOBILE PAYMENT SOLUTION	GEOGRAPHICAL COVERAGE (AS OF LATE 2001)	DESCRIPTION	ADOPTION LEVEL	PROVIDER	COMMENTS
PayDirect!	USA	Virtual POS and P2P		Yahoo! in partnership with Canadian Imperial Bank of Commerce and HSBC	• Funds are transferred to PayDirect! accounts from bank accounts and credit cards
PayPal	US (fixed-Internet solution available in over 30 countries)	Virtual POS and P2P	11 million fixed-Internet users. Particularly popular at auction sites. Unclear how many access the service from mobile devices.	x.com	• Email based • Merchants pay 30 cents for transactions under $15, and an additional 2.2% on larger transactions. • Offers services very similar to those of a bank
Sonera Mobile Pay	Finland, Sweden	Real POS (for example, vending machines, parking, fast food) and virtual POS in combination with Zed portal		Sonera	• Mainly dial-in service (for example, "dial-a-coke") • Charges aggregated at the end of the month on phone bill or credit/debit card bill

arrangements and that roaming clearinghouses adapt their current solutions (see the section *Roaming and Billing* in Chapter 3). The emergence of P2P payment scenarios will also add further pressure to develop interoperable solutions. Next, we provide a brief overview of ongoing mobile payment standardization efforts.

Mobile Payment Standardization Efforts

Today, several industrial forums, each representing different sets of interests, are participating in the development of mobile payment standards. The Mobey Forum, which we introduced earlier, has been aiming for the development of solutions that leverage emerging banking and electronic payment solutions such as the Secure Electronic Transaction (SET) protocol or the EMV standard. These standards, however, still appear several years away from broad adoption. The Mobile Payment Forum, launched by MasterCard, Visa International, American Express, and JCB in November 2001, is focusing even more specifically on mobile credit card payments. In contrast, the Mobile Electronic Transaction (MeT) initiative, which was founded in April 2000 by Ericsson, Nokia, and Motorola, has taken an approach that is more closely tied to WAP security (in other words, WIM, WMLScript SignText, and WTLS) and Bluetooth. Note that this approach is not incompatible with the objectives of the Mobey Forum or the Mobile Payment Forum. In fact, among its many associate members, which include mobile operators, security vendors, and financial organizations, MeT also counts the Mobey Forum. Other important initiatives include MasterCard's Global Mobile Commerce Interoperability Group (GMCIG) founded in June 2000, which also brings together a large number of mobile operators, financial organizations, content providers, and other m-commerce players; and Radicchio, an initiative that focuses more specifically on wireless PKI.

Different Mobile Payment Scenarios

MeT distinguishes between three broad classes of mobile payment scenarios:

Remote environment scenarios. These are scenarios in which the user accesses Web services and makes purchases from his personal trusted device (PTD). Examples include downloading a song on your mobile phone and paying for it, or connecting to a movie theatre's WAP site from a mobile phone and purchasing tickets.

Local (or physical) environment scenarios. These are scenarios in which the user goes shopping the old-fashioned way in an actual brick-and-mortar store, or goes to a restaurant and decides to pay the merchant using his PTD. In these scenarios, the consumer effectively uses his mobile device as a substitute for cash or an actual credit card he might otherwise have carried in his wallet. These scenarios could involve using Bluetooth to support local communication between the merchant's point-of-sale equipment and the consumer's mobile device for presentation of the bill and authorization by the user.

Home environment scenarios. Here, you are at home, using a fixed-Internet line to surf the Web. You might, for example, be shopping for a book on Amazon—which you could also very well do under a remote environment scenario. When prompted for payment, you use your PTD as a means of authentication to authorize the transaction.

MeT in Slow Motion

The following outlines a typical MeT remote environment scenario. The user browses using his WAP phone and connects to a WAP store through a secure WTLS session. The server provides authentication information, which is verified by the user's PTD. The user then proceeds to select some goods and places an order to purchase them. In return, the merchant's WAP site sends a payment contract to the user's PTD through a WMLScript SignText command, prompting the user to either accept or reject the charges. At this point, the user locally selects on his PTD among several available methods of payment, which could include a bank transfer, one or more credit cards, or some other means of payment such as a store debit card. Authorization to proceed with the payment is entered by typing the PIN corresponding to the selected method of payment. Upon verification of the PIN, the PTD signs the payment contract and returns it to the merchant site. Verification of the signature might either be performed directly by the merchant's Web server or via the payment system infrastructure, depending on the selected method of payment. Once the signature has been verified (assuming that it is), the merchant site sends the PTD a receipt, which is stored locally on the device for possible future reference.

The MeT forum has defined a number of such scenarios. At the request of financial institutions, it has also recently included a scenario that allows for the use of the Secure Electronic Transaction (SET) protocol, which Visa and MasterCard would like to see eventually replace TLS to process credit card payments. However, in contrast to storing to a digital wallet responsible for holding the certificates used by SET on the mobile device (not a viable option, given the device's limitations), MeT installs it on a server where authentication requests are forwarded for processing.

Concluding Remarks

Mobile security and payment are central to m-commerce. For mobile operators, mobile payment presents a unique opportunity to consolidate their central role in the m-commerce value chain (for example, through control of user authentication and the default WAP gateway), and increase *average revenue per user* (ARPU) through profitable revenue sharing arrangements with e-tailers and other content providers. This position of strength is being challenged by third-party organizations such as banks and other financial players that have started to set up their own gateways, plan to issue their own WIM cards, or offer their own mobile payment solutions. The introduction in WAP 2.0 of a protocol stack that supports end-to-end security and the emergence of Java-enabled mobile devices also contribute to creating an environment in which the operators' current position of strength could gradually erode.

For financial organizations, mobile payment is an important battleground as well, as they do not fancy being disintermediated by mobile operators, and view mobile payment and mobile banking as a way of providing added convenience to their customers along with an opportunity to reduce their operating costs. A number of solutions also involve alliances between operators and financial organizations.

Today, a number of competing mobile payment solutions have already found their way into the marketplace. Issues of standardization and interoperability will, however, have to be resolved for these solutions to reach their full potential, especially in places like Europe, home to around 60 different mobile operators whose users continually roam from one network to the other.

PART THREE

M-Commerce Services Today and Tomorrow

CHAPTER 6

Mobile Commerce Services Today

Introduction

As they follow us around, mobile devices extend the reach of fixed Internet applications and services, from accessing email and enterprise databases to trading stock or purchasing movie tickets. Beyond this, they also open the door to a slew of new services and usage scenarios, which would be inconceivable or significantly less appealing if accessed from our desktop PCs. The mobile Internet experience is very different from the one offered by its fixed counterpart. It reflects the limitations of mobile devices in the form of small keypads, tiny screens, or low bandwidth, and the demands of goal-oriented usage scenarios often involving minimal user attention. Few people would contemplate surfing the Web from a mobile phone with a five-line black-and-white screen. Yet, many can see the use for a device capable of effectively pointing you to the nearest gas station as you are fighting busy traffic. Building services capable of providing us with timely and to-the-point information while keeping user input to a minimum requires significantly higher levels of personalization and context-awareness than found today on the fixed Internet. Mobile handsets and PDAs have shown, however, that they can be more than simple productivity-enhancing tools or intelligent context-sensitive assistants. They can also help us turn "dead time" into enjoyable time by giving us access to games and infotainment services while we wait at subway stations and airports. Usage statistics from mobile portals such as DoCoMo's i-Mode (see Figure 6.1) suggest that, after email and messaging, entertainment is currently their most popular category of services, reflecting in part the young age of their user population. Figure 6.1, which compares the number of i-Mode sites and number of hits by category of services, also shows that it is not always easy to predict which service will actually prove the most popular. While in September of 2000 41 percent of all i-Mode sites were finance and insurance related, they only accounted for 4 percent of all hits. On the other hand, entertainment with 31 percent of all sites claimed 64 percent of all accesses. Location-based directories and services also represented a mere 2 percent of hits recorded by the portal, reflecting in part the fact that, at the time, these services required users to manually input their location—in 2001, i-Mode introduced position-tracking functionality with its i-Area service. More generally, as faster data rates, fancier mobile devices, stronger security, or location-sensitive technologies become more widespread, different categories of services are expected to see significantly higher usage levels—for example, location-based services and later mobile payment services. In this chapter, we take a closer look at m-commerce services available today and revisit some of the business models introduced in Chapter 2, "A First Look at the Broader M-Commerce Value Chain." Clearly, given the variety of services that have sprouted over the past few years, it is impossible to provide a comprehensive overview. Instead, we will try to give you a sense of the breadth of both consumer and enterprise applications and services found today on the mobile Internet. As we will see, many of today's services remain limited when it comes to personalization and context-awareness (including positioning). In Chapter 7, "Next-Generation M-Commerce," we revisit these issues and take a closer look at the more highly personalized, location-sensitive, and context-aware categories of services that will likely represent the second wave of m-commerce.

Here, we successively review:

- Mobile portals, including voice portals
- Mobile information services
- Mobile directory services
- Mobile banking and trading
- Mobile shopping and mobile ticketing

- Mobile entertainment
- Mobile enterprise applications and services

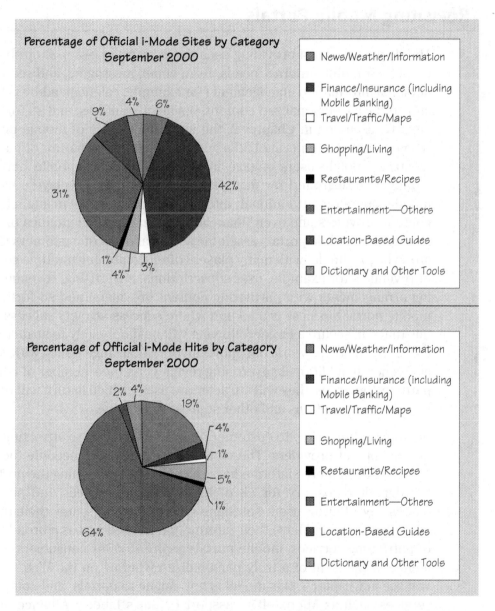

Figure 6.1 Percentage of official i-Mode sites and hits by category—September 2000. While 42 percent of the sites were finance/insurance services, they only represented 4 percent of the hits. Entertainment with 31 percent of the sites recorded 64 percent of all hits.

Source: NTT DoCoMo.

Note that we do not discuss mobile advertising, since the topic was already covered quite extensively in Chapter 2.

Revisiting Mobile Portals

Mobile portals aim at providing consumers with a one-stop shop solution to all their mobile Internet needs, from email, messaging, and search, to personal information management (for example, calendar, address book, and so forth), personalized content, payment solutions, and so forth. As already discussed in Chapter 2, the input limitations of mobile devices when it comes to entering URLs and the time-critical nature of many of the tasks mobile users engage in combine to make mobile portals a strategic battleground for many mobile players. These players include mobile operators, traditional Internet portals, device manufacturers, content providers and even financial organizations. Economies of scale require that mobile portals assemble a critical mass of customers if they are to be profitable, with many players often combining multiple sources of revenue—for example, user subscriptions, advertising, revenue-sharing arrangements with partnering content providers, and so forth. The mobile portal business is also one where success attracts success. Portals that can secure a larger following will be more likely to attract more content providers, work out more favorable arrangements with them, and generate additional advertising revenue. As their number of content partners increases, they will attract more users, which in turn will further increase subscriptions and other sources of revenue.

Personalization is key to customer retention as well as to attracting partnering content providers. The name of the game is to become the consumer's m-commerce intermediary of choice by positioning the portal as the default repository for all of his or her preferences and personal details. By offering users solutions where they can enter their preferences just once and have their information re-used across a broad range of partnering services, mobile portals aspire to offer consumers convenience in the form of a truly personalized window on the Web. (As we will see in Chapter 7, this model is not unique to portals, and competing services such as Microsoft's Passport or Sun's Liberty Alliance aim at doing exactly the same.) By also offering customers one or more payment solutions, mobile portals can provide further convenience and possibly hope to keep a percentage of every charge or transaction initiated by the user. Clearly, just as there is a variety of payment solutions, there are also a number of variations of this model. i-Mode keeps a percentage of the charges it collects on behalf of participating content partners. Originally, Yahoo! offered its PayDirect solution for free to both fixed and mobile users, although it has indicated it would eventually keep a percentage of every transaction (2.5% plus 30 cents) (see Figure 6.2).

Figure 6.3 shows examples of personalization parameters typically collected by mobile portals. These parameters can be used to personalize the user's all-important first screen—the probability of getting users to initiate a transaction is known to quickly drop with the number of clicks/screens they need to go through. Personalization parameters can also help generate short messages/alerts in the form of news updates or promotions that are directly relevant to the user's interests. Note that with the emergence of global Internet portals such as Yahoo! or AOL, many of these parameters can be reused across different access channels (for example, digital TV, desktop PC, in-car computer, and so forth),

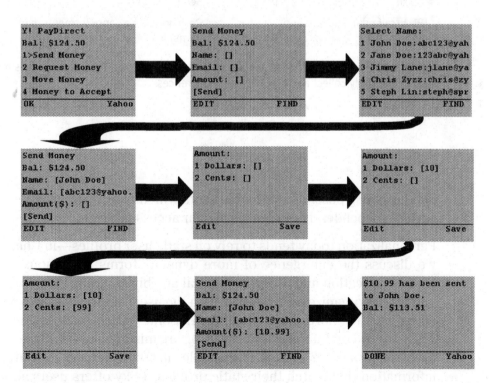

Figure 6.2 Yahoo!Mobile is an example of a mobile portal offering a payment solution to its users.

PERSONAL INFORMATION	SAMPLE USE
User Name	Login
Mobile Phone Number	To send alerts, forward emails, etc.
Email address	Customer care
Zip Code	Location-sensitive information (e.g., default weather, city guide, promotional messages)
Date of Birth	Targeted promotional messages and birthday greetings
Gender	Targeted promotional messages
List of Friends/Pals	Instant messaging, location information
Interests (e.g., sports, politics, finance)	Targeted news alerts and first screen customization
Credit Card Details or Other Method of Payment	Default method(s) of payment
Willing to Receive Promotional Messages (Y/N)?	Self evident
Maximum Number of Alerts per Day	Self evident
PIM (calendar, address book, shopping list)	Reminders (e.g., appointments, shopping items), etc.

Figure 6.3 Typical preferences and personal details collected by mobile portals today—most portals only collect a subset of these preferences.

making it also possible to provide users with a consistent Internet experience independently of their particular access platform.

Personalization today tends to rely on static user profiles—in Chapter 7, we discuss the emergence of more dynamic forms of personalization based on location and other contextual attributes. Using these profiles, portals can organize services based on the user's interests (for example, news, weather, portfolio, and so forth). Some portals such as J-Phone's J-Sky try to combine multiple ways of organizing content. Besides the usual "What's New?" option that enables users to connect to sources of information that match their static profiles, J-Sky offers users an alternative way of accessing services based on the different activities they will engage in at different times of day. Specifically, J-Sky users can select from a list of service categories such as:

- Before leaving home.
- Going to school or the office.
- Killing time.
- Shopping.
- Night life.
- Information resources for businessmen/businesswomen—two separate categories.
- Making reservations—air, rental cars, hotels, movies.

An obvious way in which J-Sky users as well as most other mobile portal users can further personalize their mobile Internet experience is through the use of bookmarks.

Voice Portals

Personalization and context-awareness are not the only way to overcome the input/output limitations of mobile devices. A complementary approach involves the use of voice both as input and output. Speech recognition and synthesis circumvent the limitations imposed by the tiny keypads of many mobile devices, and offer a completely hands-free, eyes-off-the-screen solution, which is particularly attractive for drivers, people who attempt to access m-commerce services while engaged in other activities, as well as visually challenged users. Examples of voice portals include TellMe, HeyAnita, AOLbyPhone, and Yahoo! by Phone. These portals allow users to access customized content such as stock quotes, news, weather and traffic information, horoscopes, and so forth. They also generally offer email reader services, enabling users to have their email read to them over the telephone. Some such as AOLbyPhone—the result of an acquisition by AOL of Carnegie Mellon startup Quack.com—also allow customers to shop online using voice-activated commands. These portals rely on VoiceXML, a markup language adopted by W3C since October 2001 and originally developed by the VoiceXML Forum, a consortium founded by Motorola, AT&T, and Lucent.

Mobile Information Services

With 19 percent of all hits recorded by the i-Mode portal, news, weather, and information services come in second as the most popular category of mobile Internet services. Content can be accessed both in push and pull modes. The number of information services with a mobile presence is quite daunting. Well-known players include the likes of AOL Time Warner, Bloomberg, CBS, *The Wall Street Journal*, Reuters, and many more. There are also a number of pure play niche players such as 7am.com that focus more specifically on the fixed and mobile Internet markets. Here again, personalization is key, especially in the context of push services that send regular alerts to users based on their interest profiles. The Short Message System (SMS) is often used as a delivery mechanism for news alerts, although WAP also supports push functionality. SMS has also been used to support pull interactions, with users sending SMS queries in the form of keywords to an SMS center, which returns the requested information via another SMS message. This process can also be used to subscribe to news services that will send you daily weather reports or stock quotes. Both push and pull modes of access to news and other information services have contributed to the success of SMS in Europe and other parts of the world, with a number of European operators deriving a sizable percentage of their revenues from SMS. For this business model to work, the mobile operator needs a billing infrastructure that makes it possible to charge users a premium for SMS messages they receive *reverse billing*. The operator also needs to be able to keep track of revenues generated by different content providers to flexibly and accurately implement different revenue-sharing arrangements. The lack of adequate billing infrastructure at many European mobile operators has often been cited as a major impediment to their ability to implement revenue-sharing arrangements similar to those of i-Mode, although this situation seems to be fast changing now.

Mobile Directory Services

Mobile usage scenarios often involve goal-oriented activities such as looking for services or places. Given the input/output limitations of mobile devices, keyword-based searches that return several hundred hits are not a viable option to support these activities—typing in keywords is something that most mobile users are not willing to put up with, and scrolling through hundreds of links on a five-line screen is sure to deter even the most courageous among them. Thanks to mobile directory services, this situation can easily be remedied. Directory services such as go2.com allow users to select among a broad range of categories such as restaurants, movie theaters, gas stations, ATMs, dry cleaners, doctors, and so forth. By typing in a current location—in the form of a zip code, for example—users can conveniently obtain lists of nearby places. They can also request driving directions or click through to the Web site of their choice to reserve tickets or access other relevant content.

Mobile directory services derive their revenues from several possible sources. Some charge for access, either through subscription fees or based on actual usage. Revenue-sharing arrangements with mobile operators or other content aggregators are also often part of the picture, as is advertising. Advertising is often in the form of placement fees, as is the case with go2.com, which charges merchants a monthly fee for being listed. Merchants willing to pay more can generally expect more prominent placement in the directory.

Mobile Banking and Trading

Many early m-commerce services have focused on mobile banking and other financial applications. As already pointed out in Figure 6.1, as of September 2000, over 40 percent of all official i-Mode sites were mobile finance and insurance services. Services such as Nordea's WAP Solo (see Figure 1.4) enable their customers to check their balances, review statements, pay bills, wire money, trade shares, and make online purchases at merchant sites participating in the *Solo Market* shopping mall. Banks, as we saw in Chapter 5, "Mobile Security and Payment," are also developing mobile payment solutions that aim at providing added convenience to their customers and at possibly circumventing attempts by mobile operators to position themselves as default mobile payment providers through their existing billing relationship with the customer. A number of mobile payment solutions are actually emerging in the context of strategic alliances between financial organizations and mobile operators. MobiPay, for example, started as an alliance between Spanish bank BBVA and operator Telefónica Móviles (under the name of Movilpago), and later added several other mobile operators (Airtel and Amena) as well as several banks (for example, Banco Santander, Caja Madrid, Banco Popular, and so forth) to its consortium.

Thousands of banks in the United States, Europe, and Asia already have well-established online banking solutions—the simple prospect of eliminating tens of billions of paper checks, which could translate into billions of dollars in savings, is hard to resist. It is estimated that over 10 percent of U.S. bank accounts are already online, with countries such as Canada, Germany, or Sweden posting even larger percentages. Many banks are now moving ahead with plans aimed at leveraging their online solutions to provide mobile access to their customers. Analyst firms such as Datamonitor or Meridien Research have predicted that by 2004, between 15 and 20 million Europeans will be accessing mobile banking and financial services. As they work toward providing mobile access to their online services, banks are generally pursuing three objectives:

User convenience. Offer added convenience to their customers by enabling them to access online services while on the move, and provide them with new modes of payment, which one day could end up replacing their physical wallets and the cash and various credit and debit cards they contain.

Lower operational costs. Reduce operating costs by allowing their users to access banking services that involve minimal human intervention.

Response to the mobile operator threat. Respond to the potential threat of mobile operators acting as default mobile payment providers through their existing authentication and billing infrastructure.

For financial organizations such as Nordea, mobile banking is a natural extension to efforts aimed at providing users with as many access channels as possible, an effort they trace back to 1982 when they first allowed their customers to access their bank accounts from fixed phones (see Figure 6.4). Over the years, these channels have been complemented with fixed Internet access, mobile phone access over GSM (since 1992), Internet TV access, and since 1999, access from WAP-enabled phones. Nordea WAP Solo places a particular emphasis on providing its customers with a consistent experience, independently of the access channel they use. This means same passwords, same service logics, same set of menus, although actual presentation will vary depending on whether you access your bank account from your desktop PC or your

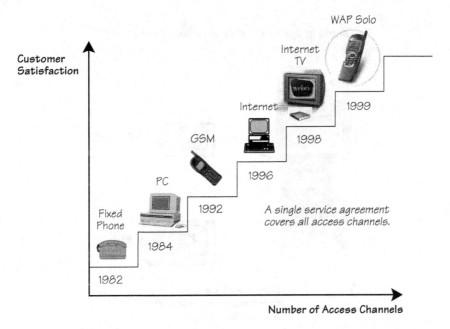

Figure 6.4 Nordea's Solo banking service: creating value by providing users with a consistent experience through as many access channels as possible.

WAP phone. Today, Nordea has over 2 million users across Finland, Sweden, Denmark, and Norway, accessing its online banking services on a regular basis, with services expected to be offered in Poland, Estonia, and other Baltic countries.

The WAP Solo service gives mobile users access to nearly all regular banking services such as checking balances, making payments, buying and selling shares, shopping, and applying for loans and life insurance policies. It also includes e-salary services where physical payslips are replaced by electronic ones, and e-billing services, where customers are invoiced over their mobile phones and can directly authorize payment without having to retype all the details. Most recently, the bank has also embarked on an Electronic Mobile Payment Service (EMPS) pilot in partnership with Nokia and Visa International. The objective is to initially support mobile payment over the Internet, and later also provide for a local merchant point-of-sale solution. The pilot relies on dual chip WAP phones with a separate plug-in size WIM card based on the EMV standard (see Figure 6.5). It will also involve the use of infrared or Bluetooth technology for POS solutions.

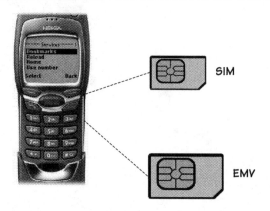

All cards in one chip inside your Wap-phone

SIM

EMV

Debit/Credit card, bank log-on,
club membership, application
downloading, etc.

Figure 6.5 Nordea's EMPS dual-chip pilot in partnership with Nokia and Visa aims at providing a payment solution that works over the Internet and can also serve as point-of-sale payment solution.

Besides Nordea, other early movers in the mobile banking sector include Okobank in Finland, which started offering GSM banking in 1996 and upgraded to a WAP service in 1999, and BTCellnet/Barclaycard, which has been offering mobile banking services to its customers since 1995, initially through SMS and more recently using WAP.

As they launch mobile services, banks can purchase their own mobile gateways and offer their services irrespective of the mobile operators their customers use, or they can partner with one or more operators— the approach taken by BTCellnet/Barclaycard.

Mobile E-Tailing and E-Ticketing

While the idea that online shopping revenue would quickly eclipse that of brick-and-mortar stores has generally proved a fantasy, (fixed Internet), e-commerce has made significant headway in a number of specific sectors. In 2000, global Business-to-Consumer (B2C) revenues reached $40 billion, twice their 1999 levels. Gartner, for example, recently reported that half of U.S. travelers used the Internet for air travel arrangements, with 30 percent booking travel online and the remaining 20 percent using the Internet for research. As with the fixed Internet, mobile e-tailing is more likely to succeed in specific niche markets such as travel (for example, air travel, rental cars, hotels) and local transportation (for example, bus, subway, taxis), ticketing (for example, movies, concerts, and so forth), last-minute purchases (for example, flowers, chocolates), food delivery, books, and CDs. At the same time, because mobile devices are carried around by their owners, they also entail new shopping scenarios. As we saw, consumers in a number of European countries, Japan, or Hong Kong are already starting to use their phones as point-of-sale payment solutions, whether to pay a merchant, board a bus, or buy a soda from a vending machine. Tomorrow, they could also use their mobile devices to compare prices at nearby merchants before making a purchase.

Mobile e-tailing services have already been launched by a number of online retailers. Sonera Mobile Pay (SMP) operated by Sonera Smart-Trust, a subsidiary of mobile operator Sonera, allows mobile users to book tickets or even buy clothes. Amazon has made its entire selection of books, CDs, and other products available to mobile customers through partnerships with a number of mobile operators and device manufacturers such as Palm, Motorola, and Nokia. The resulting Amazon.com Anywhere site takes advantage of its one-click service, which minimizes the information customers need to enter by remembering their shipping address, email, and credit card details. In 2000, Coca-Cola indicated that it would invest $100 million as part of a five-year program aimed at upgrading a number of its vending machines to support the purchase of soft drinks from mobile phones. This plan is the result of an initial *Dial a Coke* trial conducted at the Helsinki Vantaa airport in partnership with mobile operator Sonera. In the trial, Finnish customers purchased Cokes by dialing a special number (displayed on the vending machine) from their GSM phones, with the charge showing up later on their phone bills. The trial demonstrated the merits of saving people the trouble of looking for the correct change, with 30 percent of all purchases being made from mobile phones. In Japan, the vending machines are part of a new service called Cmode, where they are outfitted with liquid crystal displays, printers, and speakers to allow users not just to buy soft drinks, but to also download ringtones and screen savers on their i-Mode phones. The service is provided in partnership with DoCoMo and Itochu. Mastushita is also reported to have developed a similar service to sell alcoholic beverages and cigarettes. The service will check the customer's age, using his or her i-Mode registration information.

In general, mobile shopping offers retailers several major benefits. It provides a new marketing and distribution channel, through which they can hope to reach new customers and deliver added convenience to existing ones. Convenience can be in the form of saving them the trouble of getting the right change for their soft drink, allowing them to book movie tickets while on the go, send flowers and other last minute gifts, or purchase a book while waiting at a subway station. It could also take the form of ordering a taxi and using positioning technology to tell the taxi where to pick you up. Beyond convenience, mobile access to comparison shopping agents offers customers new savings opportunities. Examples of comparison shopping engines already accessible from mobile phones today include BookBrain, Checkaprice, DealTime, Metaprices, mySimon, PriceGrabber, and several others. As a distribution channel, mobile e-tailing also makes it possible to directly download goods such as ringtones, songs, screensavers, and maps on the mobile device for direct consumption. The mobile device can also be used to store proofs of purchase such as movie tickets, which can later be displayed or beamed when entering the theater. Similarly, when booking a trip, whether online or not, mobile devices can be used to store tickets, which can later be presented when checking in at the airport—possibly using Bluetooth. They also enable travelers to be notified of changes to their itineraries, including flight delays or cancellations, and, if necessary, to rebook flights or hotels on the go (see Figure 6.6) a particularly appealing functionality considering that 40 percent of all business bookings get modified for one reason or another.

Above all, mobile e-tailing is also about presenting customers with offers that are directly relevant to their current location or other contextual attributes, sending them promotions for nearby restaurants, movie theaters, hotels, and so forth. Clearly, this has to be done while respecting the customer's desire for privacy (for example, through pull interactions or permission marketing). These scenarios make mobile devices an ideal marketing channel, offering unprecedented personalization opportunities.

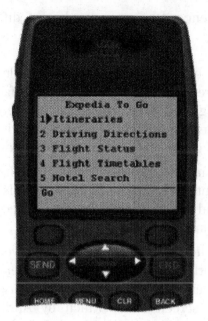 

Figure 6.6 Beyond booking flights, cars, and hotels, Expedia To Go offers customers mobile access to a number of convenient services such as flight status information or driving directions.

Finally, one should not forget that mobile e-tailing is also a great way to reduce the retailer's operating costs by minimizing human intervention in the marketing, sales, and delivery of goods and services.

Mobile Entertainment

Judging from the i-Mode statistics discussed earlier in this chapter as well as recent analyst reports, mobile entertainment will likely prove to be one of the most successful and profitable areas of early m-commerce. With close to $1 billion in total revenue in 2000, mobile entertainment has already shown that it is a perfect match for the young demographics of early mobile Internet adopters in Europe and Japan, young professionals and teenagers who have grown up surrounded by game technologies. In December 2000, 30 percent of all i-Mode users were under 25 and 60 percent under 35. Research firm Datamonitor recently predicted that by 2005, there will be 440 million mobile game players, generating over $15 billion in revenue.

Mobile entertainment generally encompasses a number of different categories:

- **Games.** There are a number of different games accessible from mobile devices. Some come pre-loaded by the device manufacturer (for example, Nokia's Snake game), and others can be accessed online (for example, WAP games) or downloaded on the device (for example, Java games or BREW games in the case of CDMA devices).

- **Music, videos, karaoke.** From the very beginning, i-Mode has featured karaoke sites. More recently, a number of mobile portals allow users to listen to and download songs on their mobile devices, and several MP3 phones have reached the market.

- **Screen savers, ringtones, caller group icons.** Screen savers and ringtones have proved particularly popular among mobile users, who generally relish personalizing their devices. We mentioned earlier the success of Bandai's Chara-Pa screensavers, which by April 2000 boasted 1.6 million users, each paying the equivalent of one dollar a month for the privilege of downloading a new screensaver each day.

- **Jokes and horoscopes.** Yet another way of killing *dead time* while waiting at airport lounges and train stations.

- **Digital postcards.** Users can send each other digital postcards and photos using Multimedia Messaging Services (MMS).

- **In-car telematics entertainment.** This can prove rather useful when it comes to keeping your children busy in the back seat.

- **Dating services.** Yet another category of mobile entertainment services set to appeal to those many mobile users 35 and under.

- **Gambling.** While only legal in some countries and states, gambling is known for being a particularly lucrative business and one that fits well with the impulse, spur-of-the-moment nature of many mobile usage scenarios. Reuters Insight reports that licensed gambler operators have profit margins of around 50 percent, bookmakers' margins ranging between 20 and 35 percent, and distribution channels for licensed gambling operators can achieve margins ranging between 6 and 17 percent. In the United States alone, it is estimated that there are 1 million daily online (fixed Internet) gamblers.

Statistics from Nokia indicate that mobile games are generally played while waiting or when at home or in cars—presumably not in the driver's seat (see Figure 6.7). Early mobile games have been primarily single-player, which often came pre-loaded on the mobile device. More recently, however, a number of new offerings have hit the market, first relying on SMS messages and more recently on WAP and even Java. A number of new games are multiplayer games, and some even involve location-sensitive features. For example, BlueFactory, a popular mobile game publisher, offers a game, *Hunters and Collectors*, where players use SMS messages to shoot at each other when they are within proximity. In general, mobile games can give rise to a number of different sources of revenue:

- **Traffic-based revenue.** For mobile operators, games present an important source of data traffic revenue, whether through SMS messages, access to WAP servers, or through traffic generated from downloading games on mobile devices. Additional traffic can also be generated by offering community chat rooms where players can exchange tips, or by generating event notification messages for the organization of multiplayer games.

- **Usage charges.** As with other sources of content, players can also be billed on a per-play basis or through subscription fees. Multiplayer games also require player match-up services for which users can possibly be charged extra.

- **Advertising and sponsorship.** Advertising and sponsorship are yet another typical source of additional revenue and can include prizes for contests.

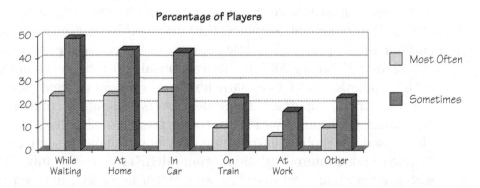

Figure 6.7 Nokia mobile game users play mostly while waiting or when at home or in the car.

Digital Bridges: A Typical Game ASP

As already discussed in Chapter 2, delivery of content such as games and entertainment often entails partnerships among several players such as mobile operators, game publishers, mobile portals, advertisers, and even platform manufacturers (for example, manufacturers Alcatel and Trium pre-load game publisher In-Visio's ExEn virtual machine on some of their mobile phones). Revenue-sharing agreements therefore need to be worked out, and many mobile game publishers operate as wireless ASPs rather than attempt to interact directly with the customer. A typical example of this model is Scottish game ASP Digital Bridges. The company essentially relies on a two-pronged approach to business partnerships. For distribution, it partners with mobile operators and mobile portals—as of late 2001, the company has entered into over 20 such partnerships with the likes of Genie and SprintPCS. On the development front, Digital Bridges offers content partners a platform-independent environment, UNITY, in which they can develop games that will run on a broad range of devices and networks. Digital Bridges takes care of distributing the games through its many partnership arrangements with mobile operators and mobile portals. UNITY works on devices running WML, HDML, SMS, and cHTML, and uses profiles to tailor games to the capabilities of different handsets. The platform also includes detailed usage monitoring functionality, enabling sophisticated revenue sharing arrangements with partnering mobile operators, and mobile portals based on transactional and subscription revenues as well as airtime charges.

Standardizing revenue-sharing arrangements and reverse billing and developing standards for cross-operator delivery (for example, roaming scenarios) are all central to the broader take-up of mobile entertainment—as well as mobile information services. Consortia such as the Mobile Entertainment Forum (MEF) founded by a group of mobile entertainment players, mobile operators, and mobile technology providers in February 2001 are actively pushing for the development of such standards. As of late 2001, however, only 50 percent of the 60 European mobile operators supported reverse billing for content delivery—whether through WAP or through premium SMS messages still often used by mobile information and entertainment services. The Mobile Gaming Interoperability Forum (MGIF) founded in July 2001 by mobile device manufacturers Ericsson, Motorola, Nokia, and Siemens is also attempting to define standards that will facilitate the development of games capable of running across a broad range of mobile networks and mobile devices. Other consortium members include game publishers and developers, game platform vendors, game service providers, and service integrators, many of which are also members of MEF.

Mobile Business Applications and Services

Just as a number of consumer services benefit from being extended to the mobile world (for example, ticketing, news, banking, entertainment), so do enterprise applications and services. The result is a number of new usage scenarios and the emergence of applications and services that only make sense in a mobile context. Many of these solutions are being adopted as part of initiatives to *mobilize the workforce*, enabling employees to remain in touch with their email, collaborate with colleagues back at the office, and access critical company databases and enterprise systems while on the move. As a result, major enterprise solution vendors such as SAP, Oracle, Siebel, IBM, Sybase, and many more have all developed mobile extensions to their offerings—some capable of being accessed from mobile phones and many others developed with PDAs, palmtops, or even laptops in mind. Examples of mobile enterprise applications include:

Mobile Enterprise Resource Planning (m-ERP) and mobile Supply Chain Management (m-SCM). Enterprise Resource Planning (ERP) systems support procurement, production, and distribution processes, keeping track of customer orders, inventory levels, Bills of Materials, and so forth. By enabling their sales force to type orders straight into the ERP while at a client's site, companies can increase employee productivity and reduce clerical mistakes. By allowing them to check production schedules and inventory levels, and access all important product configuration and Available-To-Promise/Capacity-To-Promise (ATP/CTP) functionality to obtain real-time delivery quotes, they empower their sales force to make more competitive and realistic offers to customers. Today's ERP systems tie into broader supply chain management solutions that extend visibility across multiple tiers in the supply chain. Mobile Supply Chain Management (m-SCM) is about empowering the workforce to leverage these broader systems through similar inventory management and ATP/CTP functionality that extend across multiple supply chain partners and take into account logistics considerations.

Mobile Customer Relationship Management (m-CRM). Customer Relationship Management (CRM) is about enhancing interactions between an enterprise and its customers, whether before, during, or after the sale. CRM systems help keep track of prior interactions with the customer, what their preferences are, what products and services they have purchased, what maintenance they have required in the past, and so forth. As such, CRM solutions cover a broad range of enterprise activities, many of which can benefit from mobile access with some scenarios overlapping with the m-ERP/m-SCM *sales force support* scenario outlined previously. For example, by remotely accessing a customer's order history, a salesperson is in a better position to identify cross-selling or up-selling opportunities. By enabling its *field repair* people to remotely access a customer's prior maintenance and repair records, it is likely to also enhance the productivity of its field maintenance workforce and increase customer satisfaction. In addition, mobile field repair solutions also help streamline and enhance the accuracy of the billing process and can help generate sales leads.

Mobile health care solutions have been deployed by hospitals to help physicians and nurses remotely access and update patient records; here again, improving productivity, reducing administrative overhead, and enhancing overall service quality.

Mobile telemetry solutions can also offer significant cost-saving opportunities, enabling technicians to remotely access usage statistics and possibly perform preventive maintenance operations. In Finland, such applications have been deployed to remotely monitor and control water pumping stations. Closer to home, a number of utility companies are contemplating deployment of mobile telemetry solutions that would enable them to access usage information without having to enter private homes. Car manufacturers are also studying the use of mobile telemetry applications for remote vehicle diagnosis and preventive maintenance.

Mobile fleet tracking and dispatching can help companies manage fleets of vehicles (for example, delivery vans, field service workforce, security patrols, tow trucks or taxis), tracking their position and dispatching them to new locations. These solutions help improve productivity and reduce response time—hence also improving customer satisfaction.

Many of the above scenarios empower employees through mobile access to new or existing enterprise applications. As such, they do not really qualify as m-commerce services, since they do not directly entail transactions between a service provider and a customer. Over time, however, it is expected that B2B services such as exchanges and e-marketplaces will also offer mobile extensions. This is not to say that m-commerce does not already have a strong B2B dimension today already. To the contrary, as we have seen throughout this book, most m-commerce value chains tend to involve a number of Business-to-Business transactions with many players positioning themselves as ASPs that share revenue with mobile operators, mobile portals, content providers and so forth.

As a final note, while the success of many early mobile Internet services such as i-Mode have been based on a rather young user population, the enterprise market for mobile services and applications should not be discounted. In fact, it will likely be crucial to the early uptake of 2.5G and 3G services. It is not by accident that mobile operator BTCellNet first targeted corporate users when it launched its GPRS service.

Concluding Remarks

In this chapter, we reviewed some of the most prominent m-commerce services and applications today. They include simple news services, directory services as well as mobile e-tailing, ticketing, banking, and entertainment services. The early days of m-commerce have been marked by the success of relatively simple infotainment services and the dominance of mobile operators who tightly control access to the customer through their billing and authentication infrastructures. Over time, other services and applications will grow in popularity by offering added convenience to customers and enhanced productivity to enterprises and their employees. By itself, mobile entertainment is already generating $1 billion in annual revenue and is set to explode in the years to come. The mobile portal market is estimated to grow to between $5 billion and $10 billion by 2005—with many analysts projecting revenues as much as five to 10 times this figure for m-commerce as a whole.

With the emergence of finer location-tracking functionality, more sophisticated approaches to personalization, and new standards aimed at supporting automatic service discovery access and composition, another wave of more sophisticated services seems set to emerge in the years to come. These new solutions will likely force mobile operators and mobile portals to rethink their position in the value chain and are the topic of our next chapter.

CHAPTER 7

Next-Generation M-Commerce

Introduction

Today's m-commerce landscape is dominated by relatively simple info-tainment services, as illustrated by the i-Mode usage statistics we discussed in Chapter 6, "Mobile Commerce Services Today." Personalization is generally limited to a small set of static preferences that do not reflect the changing context within which a user operates. In practice, one's interest in a particular service such as the weather in Pittsburgh or traffic conditions to the airport is tied to a particular context—being in Pittsburgh and having a plane to catch in two hours. The needs and preferences of users change all the time with the activities they engage in, their location, the people they are with, and so forth. Capturing this broader context while minimizing the amount of input required from the user is central to creating a more compelling m-commerce experience. A second key element in the emergence of what we will refer to as the *next wave of m-commerce services* is the development of applications that can intelligently locate and take advantage of multiple services. Think of a travel concierge application, capable of booking a flight, then connecting to a rental car site to reserve a car at the time a plane is scheduled to arrive at its destination. Developing such applications requires languages to describe services, where to find them, what capabilities they support, how they can be invoked, and so forth. As mobile services become increasingly aware of our context, they also open the door to new marketing and advertising opportunities such as sending us coupons around lunchtime for the sandwich place around the corner. Such scenarios raise important privacy issues and will require mobile industry players and regulators to develop solutions that strike a proper balance between users' desire for convenience and their concerns about their privacy.

The following attempts to illustrate what next-generation m-commerce services might look like a few years from now.

Next-Generation M-Commerce Scenarios

Consider the scenario depicted in Figure 7.1. John's calendar system reminds him that today is Valentine's Day. The reminder comes with a menu offering to help him buy flowers for Mary, his girlfriend, and arrange an evening out. In the simplest variation of our scenario, John first clicks on the *buy flowers* option, selects from a number of online flower shops, arranges for payment, then moves on to book two seats to see *Carmen* at the opera, reserves a table at a nearby restaurant, and arranges for transportation. While this scenario is possible using today's technology, it is hard to imagine that John would actually want to carry out all these transactions from his mobile phone—there would simply be too many screens to go through, too much information to type and possibly too many details to memorize or jot down. Thanks to a number of emerging technologies and standards, this type of scenario will become more commonplace in the years to come.

Higher degrees of personalization and interoperability. Collectively, the mobile portals and many mobile services and applications with which we interact know quite a bit about us. However, when it comes to offering users a truly personalized experience, they fall short, as this information is often spread across systems that do not know about each other or cannot talk to one another. Recently, however, several solutions have been proposed that aim at collecting user preferences and details in a single repository that can be tapped across a number of applications and services (for example, 3GPP's Personal Service Environment, Microsoft's Passport, and other *single sign-on/e-wallet* systems such as the one proposed by Sun's Liberty Alliance or AOL's Magic Carpet/Screen Name Service). These solutions, which still need to evolve and do a better job at addressing *pri-*

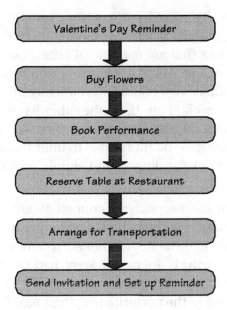

Figure 7.1 Valentine's Day scenario.

vacy issues, offer the prospect of much higher levels of personalization. In our scenario, John's Valentine's Day service would know about his girlfriend, Mary, and could automatically prepare a customized message to go with the bouquet, or fill in the address where it should be delivered. John and Mary might also have instructed their respective e-wallets to share calendar information, allowing John's service to plan the evening in such a way that it does not conflict with Mary's business commitments earlier during the day. Languages such as the *Simple Object Access Protocol (SOAP)* are already making it much easier for different applications and services to talk to one another and exchange information such as the calendar details in our example. Over time, higher degrees of interoperability will make it possible to leverage ever more disparate sources of personal information. Together with data mining and collaborative filtering techniques capable of discovering preferences not directly specified by the user, these solutions will provide for unprecedented levels of personalization.

Location-sensitive applications and services. Mobile operators around the globe have started to deploy positioning technologies that will enable them to pinpoint the location of their users. While some of these efforts are driven by security considerations (for example, E911 in the United States), this technology is also prompting the development of applications and services that are capable of taking our location into account. When John and Mary leave the restaurant and decide to call a taxi, their mobile device could automatically fill in the location where they want to be picked up. If, on the other hand, John happens to be out of town on that day, his Valentine service—possibly with the help of his calendar system—might be able to infer that there is no way he and Mary can go out for dinner, and simply arrange to send her a bouquet.

Context awareness. Beyond location sensitivity, context awareness is about capturing the broader context within which a user operates: the activity she is engaged in, the people she is with, her surrounding environment, the weather and so forth. If John and Mary had originally planned to walk from the opera to the restaurant and it suddenly started to rain just before the end of the performance, their mobile service could offer to arrange for a taxi to come pick them up. Taking into account calendar information such as Mary's business commitments earlier during the day is yet another example of context awareness.

Web Services. Maintaining directories of services such as those found today on mobile portals or those maintained by sites such as Zagat-Survey, a sophisticated online restaurant directory, can only help so much. Even with these online directories, finding the right service remains a tedious process that involves manually connecting to a number of individual sites and checking details such as opening hours or whether a flower shop offers a special Valentine bouquet. Emerging *Web Services* standards such as Universal Description, Discovery and Integration (UDDI), the Web Services Description Languages (WSDL), and ongoing research on what is generally referred to as the *Semantic Web* offer the promise of significantly simplifying this process. These advances will facilitate the automated discovery, invocation, and composition of services and lead to a significantly richer set of user functionality. Say, for example, that John's Valentine service is holding two tickets to go see *Carmen* and is now proceeding to find a restaurant; it would want to use the address of the opera house to locate a nearby restaurant, and the time when the opera is scheduled to end to determine when to book the table. Service composition is precisely about supporting scenarios that involve accessing multiple services, using information obtained from one when invoking another.

Intelligent scripts and agents. Ultimately, as our mobile computing environment becomes increasingly aware of our context, and as Web Services standards emerge, it will become possible to build more sophisticated applications and services. These could operate as intelligent scripts or specialized agents that automatically or semi-automatically combine a number of steps such as the ones outlined in Figure 7.1. On Valentine's Day, John could be presented with three or four high-level scripts from which to select, say a "Flowers and Night Out," "Flowers Only," and "Flower and Dinner at Home." Once selected, the intelligent script could proceed with minimal user input. "Flowers and Night Out" would automatically discover and access relevant flower services, identify the best site from which to order a bouquet for Mary and find the combination of opera and restaurant that best matches John and Mary's requirements. In the process, the script would require minimal input from John, just asking him to double-check some details and eventually confirming with him that he is happy with the proposed evening arrangements.

In the remainder of this chapter, we take a closer look at each of these functionalities and outline some of the new m-commerce services and business models they are giving rise to.

Personalization

One of the most compelling aspects of the Internet is its ability to support mass customization, providing each user with his or her own individual channel to a business's online presence. Personalization on the fixed Internet today combines a number of sources of information, collecting data directly from the user, using cookies to track user interactions with the Web site, and possibly attempting to uncover additional information through data mining or collaborative filtering techniques. Personalization is key to user convenience, as it helps present users with relevant content. Jupiter Communications reports that on the fixed Internet, users are 25 percent more likely to return to a Web site they have personalized. While even on the fixed Internet, personalizing a site can be a somewhat aggravating process, few mobile users would ever contemplate filling forms from their mobile devices. Instead, most mobile portals and content providers allow their users to enter their profiles from the convenience of a fixed Internet connection. Even so, this process remains extremely tedious, as each mobile player requires users to re-enter their personal details and preferences over and over again. Mobile portals, mobile operators, some banks, and a number of other players are all hoping to position themselves as the user's preferred personalization partner, offering a one-stop shop personalization solution that can be leveraged across a broad number of partnering content providers. In Chapter 6, we reviewed some of the typical information collected by mobile portals such as Yahoo!Mobile or i-Mode. Beyond today's mobile portal solutions, several other initiatives aim at providing one-stop shop personalization solutions.

3GPP's Personal Service Environment

For example, 3GPP, as part of its Open Service Access architecture (see *Evolving Application Architectures—How Open is the Mobile Internet?* in Chapter 4), is working on plans to provide a Personal Service Environment (PSE)—at the time of writing, the exact name of this environment is still the subject of discussion. The PSE would serve as repository for a common subscriber profile across all 3G services. The profile would contain personal details such as telephone number, email address, street address, and subscription information, as well as payment details and preferences that are likely to be re-usable across a number of different services. Additional information would still be collected by individual services to provide higher levels of personalization. Under the PSE model, users would only have to log on once per session. Once authenticated within their PSE environment, they would gain access to all participating services to which they subscribe. Their personal details and preferences would selectively be made available to these services, enabling them to instantaneously personalize content even on the user's first visit to a site.

Microsoft's .NET Passport

Other major contenders in the one-stop shop personalization market include Microsoft's well publicized .NET Passport initiative, and rival competitor Sun's Liberty Alliance, which features a number of key mobile players such as Nokia, Vodafone, and NTT DoCoMo.

Microsoft's .NET Passport is an ambitious initiative that ultimately aims at integrating isolated islands of personal data by:

- Offering a single sign-in service and a central repository (or e-wallet) called *Passport* for all the user's key personal information. Passport's sign-in service offers one click registration at participating sites, saving users the trouble of logging in multiple times and memorizing multiple passwords. Passport can store a user's payment details, which she can selectively re-use at participating merchant sites for easy shopping. The service could ultimately store different application settings, the user's favorite Web sites, her preferences, the devices she owns, and possibly even contextual information such as her physical location.

- Empowering users to review their personal data and selectively control who has access to them.

In the future, people could use e-wallet solutions like Passport to selectively allow different services and applications to exchange their personal information. Using languages such as the Simple Object Access Protocol (SOAP), a travel service could update your calendar with information about your next business trip and your personal restaurant concierge application could be given access to your friend's calendar when planning an evening out.

While as of late 2001 only a preliminary version of .NET Passport was available, Microsoft already claimed over 200 million users—mostly Hotmail account holders. By simplifying the sign-in process and enabling users to re-use their personal information across a broad range of services, Microsoft views its .NET initiative as a way to help move the Internet to a subscription-based business model—in contrast to the advertising-based models that characterized the early days of the Internet. As far as its own business model is concerned, the software giant expects to generate revenues from .NET through a combination of end-user subscription fees, the sale of products and services to .NET developers, and the sale of certificate-based licenses to .NET content/service providers. By making its own Windows and Office product suites compatible with .NET, the company also expects to further consolidate dominance of its traditional markets. Like efforts by global portals such as Yahoo! or AOL that aim at offering their users a consistent experience across multiple access channels, whether fixed or mobile, .NET may prove to be an important catalyst in the convergence between the fixed and mobile Internet. These efforts are all potential threats to mobile operators who, in contrast, are working to leverage their existing authentication and billing infrastructures to position themselves as the user's default personalization provider and, in the process, develop lucrative partnerships with content providers. The outcome of this competition will likely vary from one region to another. In countries such as Japan, where mobile Internet penetration is particularly high and where mobile operators have already made significant headway in positioning themselves as default payment providers, these operators and their portals are in a position of strength. In the United States, on the other hand, where the landscape is dominated by the fixed Internet, the situation is quite different, and portals such as Yahoo!, AOL, and MSN, together with initiatives such as .NET, stand a much better chance to come out on top. Europe lies somewhere in between, with the success of each mobile operator depending to a great extent on how quickly it manages to build a critical mass of content partners and establish itself as the default personalization provider with a substantial number of its users.

From a content provider's perspective, the move toward higher levels of integration of a user's personal data presents new opportunities for delivering more relevant services. This can be in the form of more personalized content as well as in the form of new advertising and marketing opportunities. By making his or her location visible to a select number of content providers, a user could be inviting location-based promotional messages. By making information about his next business trip visible to other travel services, he could open the door to new cross-selling opportunities, such as a rental car company offering him a special for the duration of his out-of-town stay. The prospect of having multiple content providers sharing cookies and mining information collected across multiple sites also raises serious privacy concerns. Microsoft has recently modified Passport to make it compliant with the W3C's Platform for Privacy Preferences (P3P). In particular, it allows users to specify which parts of their profile they are willing to share with other sites and also requires participating merchants to be P3P compliant themselves. Nevertheless, developing practical solutions that enable users to flexibly specify what information they are willing to share and for what purpose remains a daunting task, given the number of possible ways in which personal information can be used—from extremely useful scenarios to very annoying and even alarming ones. Passport currently solves this problem by limiting the number of options available to the user. Essentially, the user cannot specify that she is willing to share parts of her profile with some participating sites but not with others.

P3P on the other hand includes provisions for a language called APPEL that, in theory, gives an infinite number of options to the user but does not offer any practical way of capturing her preferences.

Location-Based Services

One of the more compelling dimensions of m-commerce has to do with the ability to track the user's position, and tailor services and promotional offers accordingly. While today most mobile operators are capable of determining the cell within which a user is located—a level of accuracy generally referred to as cell of origin (COO), more accurate positioning technologies are being deployed by a number of operators in Europe, Asia, and North America. In the United States, early deployments have been driven by the FCC's E911 requirement that mobile operators be able to pinpoint the location from which emergency calls are placed within 125 meters 67 percent of the time. While this requirement was due to take effect by October 2001, mobile operators were eventually given a one-year extension following complaints that the original deadline was overly ambitious. In Japan, since July 2001, NTT DoCoMo has been offering a location-based service called i-area that runs on top of its i-Mode service (see Table 7.1). In Europe, similar services are being offered by a number of operators such as Telia, Sonera, EMT, and Orange, and mobile portals such as Vizzavi. Besides emergency services and applications, location tracking makes it possible to support a host of lifestyle services such as finding the nearest restaurant, the nearest ATM, or the nearest gas station. Operators such as Sonera have gone one step further and offer location-based comparison shopping, enabling, for example, someone to look for the cheapest gas station in a given area. Other popular services include location-sensitive traffic and weather updates, navigation services, and location-sensitive marketing and advertising applications aimed at enticing consumers to check out local shops and restaurants. Fleet tracking, person tracking (for example, knowing where your child is), and location-sensitive billing are also often cited among the many applications and services made possible by new location-tracking technologies.

Many surveys suggest that a large percentage of mobile users are eager to have access to location-based services and willing to pay for them, leading research firms such as Ovum to predict that location-based services could generate as much as $20 billion a year in revenue by 2006.

Table 7.1 Location-Based Services Originally Offered under i-Mode's i-area Service in July 2001

I-AREA SERVICE	DESCRIPTION	PROVIDER
Weather News	Location-based weather	WeatherNews Inc.
iMapFan	Map service	Increment P. Corp.
Restaurant Guide	Restaurant guide	Pia Corp.
Traffic Guide	Traffic updates and estimated travel times	ATIS Corp.
Portable Maps	Zoomable maps with address finder function	Zenrin Co. Ltd.
Hotel Guide	Local hotel information	Narita Corp.

As indicated in Table 7.2, different classes of services have different accuracy requirements. To provide an effective directory of nearby services and stores, it is often sufficient to know the user's position within about 250 meters—which can typically be supported with COO when in a city. On the other hand, providing driving directions requires knowing the user's position within about 30 meters and recalculating this information every five seconds or so.

Table 7.2 Different Location-Based Applications and Services Have Different Accuracy Requirements

APPLICATION/ SERVICE	ACCURACY REQUIREMENTS	DESCRIPTION	POSITION REFRESH FREQUENCY
Location-sensitive billing	250m	Segmented, competitive pricing	Originated calls, received calls, mid-call
Roadside assistance	125m	Call for help	Originated calls
Mobile Yellow Pages	250m	What's near me?	Originated calls
Traffic information	Cell/Sector	What's traffic like?	Originated calls or every five minutes
Location-based messages	125m	Push advertising	Originated calls or every five minutes
Fleet tracking	30–125m	Resource management	Every five minutes or on demand
Driving directions	30m	Guidance	Every five seconds

In the following section, we briefly review available positioning and location-tracking technologies. As can be expected, higher levels of accuracy generally entail more costly solutions. Different mobile communications standards also lend themselves to the deployment of different location-tracking solutions. Several interoperability initiatives such as the Location Interoperability Forum (LIF) aim at hiding this complexity from application developers and service providers, making it possible for users to seamlessly access their favorite location-based services while roaming across networks that implement different positioning technologies.

A Brief Overview of Competing Positioning Solutions

Generally speaking, one distinguishes between handset-based positioning solutions, where the handset computes its own position, and network-based positioning solutions, where the calculations are carried out by the network.

Handset-Based Positioning Solutions

These solutions include GPS-enabled handsets, Network-Assisted GPS positioning, and Enhanced Observed Time Difference solutions. Each of these solutions is reviewed below.

Global Positioning System (GPS). GPS uses a set of satellites to locate a user's position. This system has been used in vehicle navigation systems and dedicated handheld devices for some time, and now it is making its way into mobile phones (see Figure 7.2). With GPS, the terminal gets positioning information from a number of satellites (usually three or four). This raw information can then either be processed by the terminal or sent to the network for processing, in order to generate the actual position. The U.S. government previously distorted satellite clock signals to reduce accuracy using what was called a *selective availability mask*. This practice was ended in May 2000. Consequently, GPS can now achieve accuracy between 5 and 40 meters, provided there is a clear view of the sky—when used in a car, this requires an external antenna. Chipmakers have also been able to achieve particularly high levels of integration in GPS chips, making them significantly cheaper and more power efficient than in the past. Nevertheless, GPS handsets remain relatively bulky and expensive, and suffer from very long Time To First Fix (TTFFs), essentially the

This image is about as good as we can make it. OK as is?

Figure 7.2   Benefon's GSM/GPS Esc! mobile phone.

time it takes to locate the satellites and compute the handset's position. GPS TTFFs tend to range between 20 and 45 seconds. This time can be significantly reduced using A-GPS technology.

Network Assisted GPS (A-GPS). A-GPS relies on fixed, network-based GPS receivers that are placed at regular intervals, every 200km to 400km, to help reduce TTFF computations. The fixed GPS receivers collect satellite data on a regular basis and make the information available to mobile handsets. This saves the handsets from having to decode actual satellite messages, typically reducing TTFF computations to somewhere between 1 and 8 seconds. By broadcasting their positioning data every hour or so, the fixed GPS receivers also ensure that the extra load on the mobile network remains minimal.

Enhanced Observed Time Difference (E-OTD). At the handset level, in contrast to GPS and A-GPS solutions, which both require the introduction of additional hardware, E-OTD relies only on software—a key to keeping handset prices low. E-OTD estimates the user's position by calculating the time difference that it takes to receive signals from several—typically three—nearby base stations. This technique requires that the base station positions are known and that the signals sent from these base stations are synchronized. The most common way of synchronizing the base stations is through deployment of an overlay network of *Location Measurement Units* (LMUs). The calculation itself can either be done in the terminal or on the network. In either case, E-OTD still requires handsets to have more memory, processing power, and battery power than would be needed otherwise. E-OTD offers an accuracy of around 125 meters.

Network-Based Positioning Solutions

These solutions include Cell Global Identity (CGI) and Timing Advance (TA), which are often used in combination, as well as Time Of Arrival (TOA):

Cell Global Identity (CGI) and Timing Advance (TA). CGI is the cheapest possible technique to estimate a user's position, simply relying on the identity of the cell within which the user is located. This information is available on most mobile networks today. CGI is often complemented with TA information, which essentially measures how far the user is from the base station. The resulting level of accuracy is sufficient to support location-based directory services, for example. Clearly, CGI-TA's accuracy is dependent on cell size, as well as whether the base station has an omnisector antenna or a multisector one—the latter making it possible to identify where the user is positioned in comparison to the base station. Consequently, accuracy levels will typically vary from tens of meters up to 1,000 meters in particularly large cells.

Time of Arrival (TOA). TOA uses a triangulation technique similar to E-OTD's. However, rather than relying on a downlink signal, it uses uplink signals sent by the mobile handset to three or more base stations. For this reason, it is also known as UpLink Time Of Arrival (UL-TOA). The receiving base stations, which have to be synchronized, note the time of arrival of signals from the mobile station and combine their information to triangulate the user's position. Synchronization of base stations in support of TOA is typically achieved using LMUs similar to those required by E-OTD along with GPS receivers. Because CDMA base stations are already synchronized, TOA is significantly more appealing to CDMA operators than, for example, to GSM/GPRS ones. A key advantage of TOA over E-OTD is that it relies solely on network-based calculations. In other words, it can be used with handsets that were not designed with positioning technology in mind. A variation of TOA that relies on angular measurements to triangulate the user's position is known as Angle of Arrival (AOA).

A Fragmented Landscape

Once again, the resulting landscape is a highly fragmented one, where a number of positioning solution providers such CT Motion, CPS, Qualcomm, Ericsson, Motorola, or CellPoint, to name just a few, are competing to sell different variations of the basic techniques we just outlined. As we just saw, different solutions entail different levels of accuracy and different levels of investment depending on the particular mobile communication standards on which a given operator relies (for example, CDMA versus GSM/GPRS). In October 2000, in response to this fragmented situation, Motorola, Ericsson, and Nokia launched the Location Interoperability Forum (LIF), an organization whose objective is to develop solutions that allow users to access the same location-based services independently of the particular network through which they connect and its particular positioning technology. Other related industrial efforts include the Open Location Services (OpenLS) initiative launched by the Open GIS Consortium, and the MAGIC Services initiative. MAGIC's objective is to develop a standard method for remote delivery of core geographic functionality such as:

- **Geocoding,** which is concerned with the conversion of street addresses or other human-understandable location designations into geographic coordinates (in other words, latitude and longitude).

- **Reverse geocoding,** which is the reverse process.

- **Spatial queries,** namely the location-sensitive retrieval of information, taking into account geographic proximity as well as possibly estimated travel time along a particular route.

- **Travel planning and guidance,** which include specifying travel destinations and intermediate waypoints as well as providing travelers with timely directions along the way.

In general, key challenges in promoting the broader adoption of mobile location-based services include:

- Providing developers with a single interface to access the user's location, independently of the technology used to estimate it.

- Streamlining the interface between location-based applications, content engines and databases, to facilitate the identification of content relevant to a user's location.

- Supporting solutions that facilitate roaming and ensure that users have a consistent experience independently of the operator through which they connect.

- Developing billing and revenue-sharing guidelines to facilitate partnerships between players along the location-based value chain (namely, mobile operators, mobile portals, location-based content providers, and so forth).

- Developing solutions that empower users to easily and accurately specify with whom and under which conditions they are willing to share their location information.

SignalSoft: An Example of a Location-Based Service Provider

Despite the challenges of operating in a highly fragmented marketplace and delays in deploying positioning technologies in the United States, a number of location-based service providers have emerged over the past few years. They include French company Webraska, which recently acquired Californian startup AirFlash (see Chapter 1), as well as companies such as SignalSoft, Spotcast, Cambridge Positioning Systems (CPS), TeleAtlas, and CT Motion, to mention a few. As already indicated in Chapters 1 and 2, most of these companies operate as ASPs that offer their services via mobile operators.

SignalSoft, for example, has developed a platform that enables it to capture user location information obtained through a number of different positioning technologies (see Figure 7.3), working with positioning providers such as AllenTelecom, CPS, Cell-Loc, Lucent, Ericsson, and a number of others. The core of SignalSoft's platform consists of a location manager module that acts as a gateway between the different network positioning technologies, SignalSoft's access management module, its location-based applications, and those of third-party providers. The location manager is responsible for identifying the best available location estimate in light of the requirements of different applications (for example, an emergency application versus a location-based directory application). The access management middleware, also known as *Location Studio*, is installed in the mobile operator's network. It helps control access to and billing for different subscription-based services, and allows users to specify with whom they are willing to share their location information. Location Studio also includes a development environment that enables service providers to easily build applications that blend local content, such as local restaurants, hotels or events with location information. Examples include location-based directories, local weather information, local traffic updates and location-sensitive marketing and advertising solutions. SignalSoft's own application portfolio includes

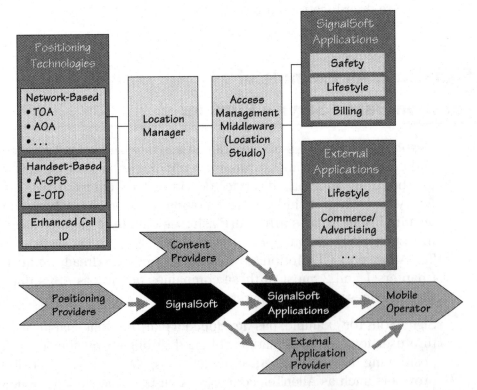

Figure 7.3 Overview of SignalSoft's location-based service platform and value chain.

E911 emergency, location-sensitive billing—enabling operators to define different geographical rate zones—and several lifestyle applications. An example is Bfound, which allows groups of subscribers to find out about each other's location. The program can be customized to support mobile workforce applications as well as to build solutions that enable your family or friends to locate you, and vice versa.

Toward Context-Aware Services

Location can be a significant differentiator when it comes to identifying relevant services. However, knowing that the user is at the corner of the street will not tell you whether he is looking for a restaurant or a laundromat. For this, we might need to know the time of day, or access our user's calendar or other relevant contextual attributes. Context-awareness is about capturing a broad range of contextual attributes to better understand what it is that the user is trying to accomplish, and what services he or she might possibly be interested in. Context-awareness is viewed by many as the Holy Grail of m-commerce, as it ultimately offers the prospect of applications that could anticipate our every wish and provide us with the exact information and services we are looking for—and also help us filter all those annoying promotional messages that we really do not care for. Few applications and services today can claim to be anywhere near this longer-term objective, although current research efforts conducted by industry and universities can help us catch a glimpse of future context-aware scenarios.

Figure 7.4 outlines the architecture of an ongoing research project at Carnegie Mellon University that aims at providing students with a context-aware environment within which they can access Internet services from their PDAs over the campus's IEEE 802.11 wireless local area network (WLAN). A context server helps keep track of the context within which each student operates. Each student's context includes his or her:

- Calendar information.
- Current location, which is regularly updated using location-tracking technology developed by local startup PanGo.
- Weather information, indicating whether it is sunny, raining, or snowing, and the current outside temperature.
- Social context information, including the student's friends and his or her teachers, classmates, and so forth.

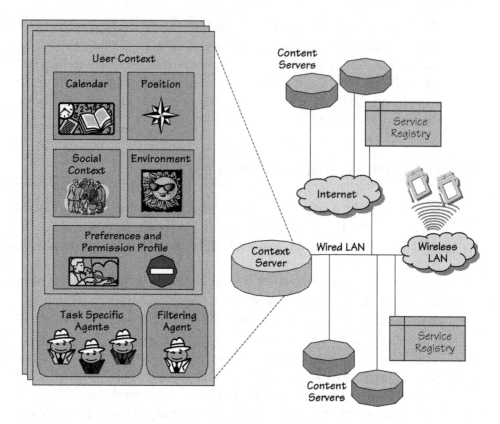

Figure 7.4 Overview of context-aware architecture for mobile Internet access over Carnegie Mellon University's wireless LAN.

This is complemented by a collection of preferences each student enters, along with a profile to help filter incoming push messages and determine who has the right to access different elements of his or her context and under which conditions (for example, "When in class, I don't want to be disrupted by promotional messages," or "If it's sunny and past 4 P.M., I'm happy to have members of my volleyball team see where I am on campus"). A user's context information can be accessed by a collection of personal agents, each in charge of assisting him or her with different tasks, while locating and invoking relevant Internet services identified through service registries. An example of a simple agent implemented at the time of writing is a *restaurant concierge* that gives suggestions to students on places where to have lunch, depending on

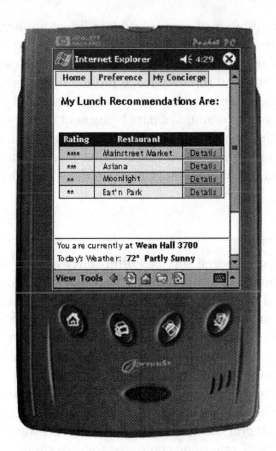

Figure 7.5 Besides food preferences, the context-aware restaurant concierge takes into account the user's location on Carnegie Mellon's campus, the weather, the user's calendar, and other contextual attributes to recommend places to have lunch.

their food preferences, the time they have available before their next meeting or class, their location on campus, and the weather (see Figure 7.5). For example, when it is raining, the agent attempts to find a place that does not require going outside of the building where the student is located. Clearly, the challenge in developing an environment such as this goes well beyond capturing objective elements of a user's context such as his calendar or location on campus. Rather, the difficulty is in deriving from contextual attributes the user's current objectives and preferences to help customize outgoing queries and filter incoming messages.

The prospect of having a number of agents and intelligent applications carrying out tasks that are potentially rather computationally intensive, such as searching for the optimal combination of airline, rental car, and hotel for a business trip, might ultimately force mobile operators, content providers, and ASPs to revise the way in which they bill users. Billing schemes that rely on the amount of data being sent over the wireless link or simple subscription models would no longer suffice. Instead, what would be needed are usage-based models that also capture the amount of processing and possibly memory used on the server side.

Concluding Remarks

While many of today's m-commerce services offer fairly limited opportunities for personalization and context-awareness, a number of new technologies have started to emerge that could potentially change the m-commerce landscape quite dramatically over the years to come. Chief among these are new positioning technologies that have now been deployed in a number of countries in Europe, Asia, and North America. Concurrently, new solutions offered or in the process of being developed by mobile operators, portals, and a number of other contenders such as Microsoft or Sun's Liberty Alliance offer the prospect of significantly higher levels of integration in our personal information. These advances in combination with emerging standards for describing, invoking, and composing Web services, as well as efforts aimed at capturing a number of elements characterizing the user's context, could lead to significantly higher levels of convenience, along with unparalleled opportunities for one-to-one marketing. This latter prospect also raises some serious privacy concerns, which will have to be properly addressed if many of the scenarios outlined in this chapter are to fully materialize. Ultimately, users will have to retain full control of their personal information and be given intuitive, easy-to-use tools that enable them to flexibly tailor access to it.

CHAPTER 8

Early Lessons and Future Prospects

Recalibrating Early Expectations

Early WAP and 3G marketing campaigns trumpeting the arrival of the Internet in everyone's pocket led many to believe that the fixed Internet experience they had grown accustomed to could simply be recreated on their mobile devices. With their tiny black-and-white screens, slow circuit-switched technologies, limited processing power, and clunky input functionality, mobile Internet devices have since forced both users and developers to adjust these early expectations (see Figure 8.1). To be sure, deployment of faster packet-switched technologies and the introduction of fancier devices will in time alleviate some of the most harrowing limitations of the mobile Internet—although at a slower pace than originally predicted. However, above all, m-commerce is not about attempting to replicate the fixed Internet experience on tiny devices. It is about adding a whole new dimension to the Web and opening the door to very different usage scenarios, where users on the move are given access to services capable of assisting them in time-critical tasks as they talk to colleagues or weave their way through busy streets. It is also about helping users stay in touch with one another and with their companies, and about helping them turn dead time into productive or enjoyable time when they would otherwise be sitting idle waiting for their next flight or traveling through the city's subway system.

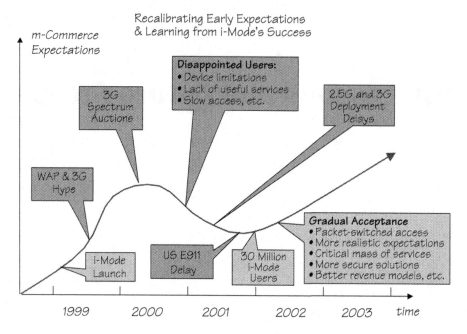

Figure 8.1 Recalibrating early expectations, and learning from i-Mode's success.

The mobile Internet landscape is populated by a growing collection of applications and services, some extending fixed Internet solutions such as messaging, online banking, entertainment, or enterprise applications, and others displaying functionality that would simply be inconceivable over the fixed Internet. Examples of the latter include mobile ticketing in combination with confirmations stored on the mobile device, location-sensitive comparison shopping to find the cheapest nearby gas station, or vending machines that enable mobile users to purchase goods from their handsets. At the core of the m-commerce value proposition is the notion that *convenience, timeliness,* and *personalization* have to make up for the clunkiness of the mobile device. To succeed, designers and developers need to rethink Internet services and move away from banner ads and flashy displays where more is better. In a mobile Internet world, where user attention and screen real estate are the most precious resources, *less is better* and information has to be to the point. Forget about search engines that return hundreds of useless hits, and focus on applications and services capable of capturing at once all the user's relevant parameters and of returning the one piece of information she is really interested in—such as the name and address of the best restaurant for her business lunch and an option to book a table and notify her client.

Early Success Ingredients

While some contrarians point to 2.5G/3G deployment delays, relaxation of the US E911 implementation deadline, or cuts in Amazon's mobile Internet efforts as signs that m-commerce might just be another flash in the cyber pan, success stories in Asia and Europe attest to the contrary. NTT Do-CoMo's 30 million i-Mode users and the billions of dollars in revenue they already generate are no accident. They mark the convergence of a number of key ingredients, which, in contrast to many other operators, DoCoMo managed to assemble almost straight from the start. Chief among them is i-Mode's authentication and billing infrastructure that enables it to assist partnering content providers in collecting fees from their users. This capability has taken much longer for many European operators to deploy, making it particularly difficult for potential content providers to develop sustainable business models—and for operators to assemble a compelling set of services. Another major element of i-Mode's success has been its packet-switched, *always-on* technology that in many ways has proved even more critical than faster data rates and that, again, has taken longer to find its way into markets such as Europe's. By providing convenience to its customers through an easy-to-use packet-switched solution that brings together a critical mass of reasonably priced services, i-Mode was able to achieve nearly instantaneous success, leaving many other operators in the dust, pondering what it would take to replicate it. This is not to say that DoCoMo did not benefit from other factors such as its dominance of the Japanese mobile communication market or low Internet penetration rates, which de facto positioned it as a cheap Internet entry solution. However, reducing i-Mode to a lucky combination of such factors would be missing the point. Today, operators worldwide are striving to emulate the Japanese operator's recipe through deployment of packet-switched technology and billing applications that finally enable them to charge for access to different services and to effectively redistribute the resulting revenues to content providers. These and other related steps will help build the more solid foundation that mobile operators, portals, and content providers need for their m-commerce services to finally take off, turning hundreds of millions of mobile phone users into as many mobile Internet consumers.

By pushing the Internet experience to new extremes—tiny screens, low data rates, and distracted users trying to carry out time-critical tasks— m-commerce also helps us uncover what it really takes to succeed in cyberspace. Gone are the days of a freewheeling industry subsidized by advertising and a skyrocketing stock market. Success stories such as those of the i-Mode m-commerce portal are precursors to future Internet business models where value is created through convenience, ease of use, and a compelling collection of content offered through no-nonsense business partnerships—content that users are actually willing to pay for.

From Killing Dead Time to Context-Awareness

The early days of m-commerce have been marked by relatively mundane services such as news, weather, traffic updates, and mobile entertainment services that, in a number of countries in Europe and Asia, have taken a young user population by storm. Even though today email and SMS messages dominate data traffic revenues, there is no killer application per se. Instead, the mobile Internet is characterized by a combination of applications and services that collectively create a compelling value proposition for a growing segment of the mobile user population. Early adopters have been a mix of teenagers, young professionals, and corporate users willing to experiment with new technologies, whether in hopes of boosting their productivity or out of social status considerations. Over time, as services such as mobile ticketing or mobile banking become more mature, reliable, and easy to use, they too will gain broader acceptance. The gradual introduction of more accurate positioning technologies will also help increase the popularity of location-based applications and services that respond to the many professional, lifestyle, and security demands of mobile users—from location-aware buddy list services, to finding nearby places of interest and requesting directions, to location-sensitive comparison shopping, location-based marketing, or location-sensitive emergency services (see Figure 8.2).

With already 1 million users in Europe alone, mobile payment could be another area poised to take off in the years to come. While key security and interoperability issues still need to be addressed, the prospect of eventually doing away with change and replacing our wallets and credit cards with more secure solutions embedded in mobile devices that we carry around anyway seems particularly hard to resist.

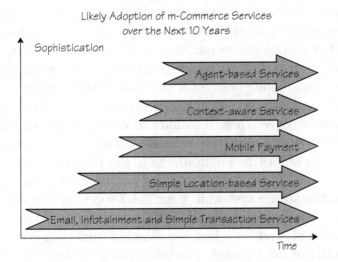

Likely Adoption of m-Commerce Services
over the Next 10 Years

Figure 8.2 From simple messaging and infotainment applications and services, m-commerce will evolve into a rich collection of sophisticated context-aware services, where users will over time benefit from semi-automated agent applications capable of carrying out tasks and accessing services on their behalf.

As discussed in Chapter 7, "Next-Generation M-Commerce," the next wave of m-commerce services will go beyond the deployment of location-sensitive services to include functionality capable of capturing the broader context within which a user operates. These services will combine multiple sources of contextual information such as calendar information, the time of year, the weather, information about other people with whom a given activity is being conducted (for example, friends, family members, colleagues), and elements of their own contexts. While such services are still several years away, they will eventually offer unprecedented levels of convenience, through functionality capable of responding to or even anticipating our needs with minimal input. Ultimately, they will open the door to agent-based functionality that will autonomously or semi-autonomously access services and carry out tasks on our behalf. For all this to happen, intuitive, easy-to-use solutions will also have to be developed that enable users to remain in control of their private information and specify who has access to it, for what purpose, and under which conditions.

Gazing farther into their crystal balls, researchers at leading companies and universities envision a *ubiquitous computing* future where an increasing number of objects in our everyday environment will offer applications and services that can be accessed from our personal mobile devices. By then, our mobile phone (or personal mobile device) might have turned into one or more wearable computers embedded in our clothes, jewelry, or wristwatches. It would serve as the ultimate remote control and would interface with a myriad of physical resources and online services, as it tailors the environment to our preferences and carries out tasks on our behalf. (Think of your mobile device all at once negotiating with the stereo system in your hotel room to find music to your liking, accessing the wall-mounted display in the room to show you the latest quarterly sales report it just downloaded from your company's intranet, and notifying room service to bring you your favorite drink.) Among other things, for this vision to materialize, interconnectivity and interoperability levels far beyond anything available today would have to be achieved, making it possible for our personal devices to automatically discover relevant objects and services and start interacting with them as we move from one place to another. Advanced roaming schemes would also have to be devised to automatically deal with authentication and access control issues, as well as revenue-sharing arrangements that such scenarios would likely entail. But, above all, significant advances would be required to support the sort of natural and intelligent dialogues any user would expect from a device that purports to carry out such a complex set of tasks on his or her behalf.

From Authentication and Billing to Personalization

M-commerce has opened the door to new entrants, while also forcing many players to re-evaluate their position in the value chain. To mobile operators in particular, m-commerce is both a major source of opportunity and a threat of being left behind as low-margin data pipes. As they attempt to follow in the footsteps of NTT DoCoMo, they need to leverage their existing authentication infrastructure and the billing relationship they have established with their customers to develop partnerships with a critical mass of content providers—positioning themselves as billing providers and/or mobile portals. In many ways, because of limitations in existing standards when it comes to supporting end-to-end security, mobile operators today are in a position of strength. As standards such as WAP evolve, and as alternative end-to-end security solutions become more widely available, this advantage will erode and force operators to seek other sources of strategic advantage. A key element in this strategy will be to position themselves as default personalization providers, holding the user's personal preferences, his or her personal information (for example, location, address book, buddy lists, and so forth), as well as his or her position information and payment details—information that users are reluctant to enter over and over again and that holds the key to delivering a convenient and timely user experience. As they try to reposition themselves as default personalization providers, mobile operators and their mobile portals will find themselves competing with global portals such as Yahoo! or AOL, and with other powerful contenders such as Microsoft's .NET initiative or Sun's Liberty Alliance. In a market already dominated by economies of scale, this situation can only further accelerate the pace of merger and acquisition activities and the development of strategic alliances.

Convergence of the Fixed and Mobile Internet

The mobile Internet marks the convergence between a world of open standards and one that has long been characterized by a mosaic of incompatible networks, and a legacy of markets long dominated by national (often state-owned) monopolies. The fragmented mobile communication landscape as well as the limitations of today's mobile devices and mobile communication standards have given rise to protocol suites such as WAP, which is set to dominate the mobile Internet landscape for the years to come. Despite its idiosyncrasies, which it derives from the many mobile communication standards it tries to support, WAP is slowly converging with IETF and W3C standards—which themselves are evolving to accommodate the demands of the mobile world. This convergence is best illustrated in WAP2.0 with the adoption of XHTML Basic and the introduction of a second protocol stack specifically intended for faster bearer services with built-in IP and capable of supporting the regular HTTP/TCP/TLS Internet protocols. Another key driver behind the convergence between the fixed and mobile Internet will be user demand for a consistent experience across different access channels. Users will expect to have their personal profiles and information reflected in all of their Internet interactions, independently of access channel. To survive, content providers and portals will have no choice but to establish a multichannel presence, whether directly or through partnerships with others. However, even in the long-run, the mobile Internet will continue to retain unique attributes through services and usage scenarios that are purely mobile in nature (for example, location-based services, mobile point-of-sale payment solutions, as well as future context-aware services yet to be developed).

Concluding Remarks

With tens of millions of mobile phone converts, the question is not whether m-commerce will ever materialize, but how quickly. The fact that i-Mode was able to ramp up from 0 to 30 million customers in less than three years suggests that when all the right ingredients are in place, m-commerce can gain rapid acceptance. Armed with a better understanding of what these ingredients are and with lessons gleaned from earlier setbacks, efforts to replicate the i-Mode success story are now well on their way in countries around the world. A number of critical interoperability efforts have also been launched that will pave the way for higher levels of personalization and context-awareness in future services. As location-based services, mobile payment solutions, and a slew of novel applications and services yet to be invented find their way into the marketplace, growing segments of the mobile phone population will find the convenience of m-commerce increasingly hard to ignore. Looking farther into the future, who could resist the prospect of having their own personal context-aware butler arranging for restaurants, hotels, and taxis for a mere $10 per month?

Glossary

2G Second-generation mobile communication systems. Examples include GSM, TDMA, PDC, and cdmaOne.

2.5G Two-and-a-half-generation mobile communication systems. Umbrella term used to designate intermediate packet-switched, *always-on* systems. Examples include GPRS and CDMA2000 1X.

3G Third-generation mobile communication systems. Key features of 3G systems, as originally recommended by the International Telecommunication Union (ITU) as part of its International Mobile telecommunications 2000 initiative (IMT-2000), include a high degree of commonality of design worldwide, worldwide roaming capability, support for a wide range of Internet and multimedia applications and services, and data rates in excess of 144 kbps. Examples of 3G standards are WCDMA/UMTS, EDGE, and CDMA2000.

3GPP/3GPP2 Third-Generation Partnership Projects in charge of laying out the specifications of 3G standards. The 3GPP partnership focuses on WCDMA/UMTS and EDGE, while 3GPP2 focuses on CDMA2000.

(intelligent) agent Program capable of carrying out tasks on behalf of a user such as consulting one or more Internet services and possibly initiating transactions with services such as booking a hotel or purchasing flowers. Simple examples of agents (or bots) include comparison-shopping engines. In the future, with the emergence of Web services, more sophisticated e-commerce and m-commerce agents are eventually expected to emerge.

A-GPS Assisted GPS.

air interface The common radio interface between the mobile station and the base station in a mobile communication network; examples include GSM or cdmaOne.

alert Short message sent to a mobile user to keep him or her updated about the news, weather, traffic conditions, and so forth.

AMPS Advanced Mobile Phone System. Analog cellular network system used in the United States.

ANSI American National Standards Institute.

API Application programming interface.

application Program that performs a specific task or function. Examples range from electronic calendars and telephone books to navigation systems and games.

APPEL A P3P Preference Exchange Language. Language under development by W3C to model user privacy requirements.

ARIB Association for Radio Industry and Business. Telecommunications standards body for Japan.

ASP Application service provider. A type of company that provides the functionality of a software application, such as billing, game, or navigation, on a remote-usage basis to corporate customers (for example, a mobile operator).

asymmetric cryptography An encryption method where a pair of keys (one public, one private) is used to encrypt and decrypt messages.

authentication The process of verifying the identities of parties involved in a communication or transaction.

authorization (credit) Operating regulations whereby a transaction is approved by an issuer (or authorized agent), before the transaction is completed by the merchant.

B2B Business-to-Business.

B2C Business-to-Consumer.

bandwidth Measures the data-carrying capacity of a communications channel—can be expressed in bits per second (bps).

banner ad Web advertisement where graphics and text are combined to entice users to click through and gain more information on a particular product, service, or organization.

base station (BS) Radio transmitter/receiver that maintains communications with a mobile telephone over the air interface. Each base station is responsible for communication with all mobile phones in a given geographical area or cell.

bearer service A telecommunications service that allows the transmission of information between a mobile station and the network. Examples include GSM, SMS, GPRS, cdmaOne, and so forth.

bill presentment Online delivery of bills to customers. The latest bill presentment solutions combine invoice delivery with online payment options.

billing Charging users for access to different services, and collecting payment.

Bluetooth Short-range wireless data communication technology. First-generation Bluetooth technology was introduced to operate in the 2.4-GHz band and send data at 720 Kbps within a 10-meter range. The second generation (BlueCore2) increases the network's range to 100 feet.

bps Bits per second.

BREW Binary Runtime Environment for Wireless. A thin, standardized execution environment developed by Qualcomm to run across all CDMA handsets. Native BREW applications are written in C/C++, although BREW also supports Java programming.

bricks-and-mortar Term used to describe traditional businesses that do not rely on an online presence to market or sell their products or services.

browser Program used to access and display information at different Web sites.

BSC Base station controller. Mobile communication network element responsible for managing multiple base stations within a given geographical area and for coordinating the handover between base stations as a mobile station moves from one cell to another.

BSS Base station subsystem.

BTS Base transceiver station—another term for a base station.

business model Combination of services, partnerships, sources of revenue, and charging structure that a company relies on to be in business.

Business-to-Business (B2B) E-commerce transactions, where both the buyer and seller are businesses.

Business-to-Consumer (B2C) E-commerce transactions where the buyer is an individual acting in a personal capacity and the seller a business.

CA Certification authority.

CA certificate A certificate for one CA issued by another CA or root.

Caller Group icons Allow users to associate pictograms with different groups of callers.

CAMEL Customized Applications for Mobile Networks Enhanced Logic. A tool that enables GSM operators to offer their subscribers operator-specific services even when they roam outside of their home network.

CDMA Code Division Multiple Access. A type of digital cellular network that allows multiple channels to coexist in space, time, and frequency by assigning them codes so that they minimally interfere with one another. Examples of CDMA standards include cdmaOne and the different variations of the CDMA2000 standard.

CDMA2000 An implementation of wideband CDMA, a 3G mobile communications standard.

CDMA2000 1X A 2.5G variation of the CDMA2000 standard (sometimes also referred to as a 3G standard, given its support of theoretical peak data rates of 300 kbps).

CDMA2000 1XEV Full-fledged 3G implementation of CDMA2000 expected to support peak data rates of up to 2.4 Mbps.

cdmaOne 2G CDMA mobile communications standard also referred to as IS-95. The initial version of the standard is known as IS-95a, while a more recent version with peak data rates of 64 kbps is known as IS-95b.

CDPD Cellular Digital Packet Data. A communication standard also known as IS-54 that adds transmission capabilities to AMPS networks.

CDR Call detail records. Call records collected by mobile operators.

cell The geographic area of radio coverage of a base station.

Cell global identity (CGI) Technique that relies on the identity of the cell within which the user is to estimate his or her position.

certificate A digital record that contains indentifying information about an entity such as a person or organization. This includes the entity's public key, an expiration date, and a certificate serial number, along with the signature of the certification authority that generated it.

Certification authority (CA) Also known as certificate authority, an entity that issues certificates for purpose of digital identification.

channel A conduit for (wireless) communication between initiating and terminating nodes.

cHTML (Compact HTML) A restricted form of HTML originally used to encode content in i-Mode. Recently, i-Mode has indicated that it is switching to XHTML Basic.

churn   The tendency of customers to switch from one mobile operator or service provider to another. A low churn rate is a sign of high customer loyalty. Higher churn rates entail higher operational costs.

cipher text   Text that has been encrypted to make the data unintelligible, as opposed to plain (or clear) text.

circuit-switched   Circuit-switched communication reserves bandwidth for the entire duration of a communication, independently of how much information is actually being transmitted at any point in time. Circuit-switched communication is well suited for voice communication, but is particularly inefficient when it comes to supporting the bursty data traffic patterns associated with Internet access.

CLDC   Connected limited device configuration.

clicks-and-mortar   Borrowed from the expression *bricks-and mortar*, it refers to companies that have successfully integrated their online and offline marketing and sales channels.

click-through   Process by which a user is transferred to a Web service by clicking on an online ad for that service.

client   An application running on a PC or mobile device and relying on a server to perform some tasks.

collaborative filtering   Technique used by companies such as Amazon to recommend products or services based on information collected from customers with similar tastes and interests.

confidentiality   Ensuring that only the sender and intended recipient of a message can read its content.

Connected limited device configuration (CLDC)   A J2ME configuration particularly well suited for mobile devices. CLDC relies on a exceptionally small virtual machine known as the K Virtual Machine (KVM), where K is used to indicate that the machine's footprint is measured in kilobytes rather than megabytes—somewhere between 40 and 80 kilobytes, depending on different compilation options and the device on which the virtual machine is installed.

content aggregator   Company that combines and/or repackages content from multiple content providers.

content provider   Generic term used to refer to any organization that creates and maintains information or services for online access.

context-aware service/application   Service or application that adapts to one or more elements of a user's context, such as his or her location, activities, the people he or she is with, the weather, and so forth.

cookie   A small file stored on a user's Internet access device (for example, computer or mobile device). Cookies are used by Web sites to recognize users when they return. They are used to keep track of their interactions with the site.

CRM   Customer relationship management. Methodology and tools used to enhance interactions between an organization and its customers, whether before, during, or after a sale. CRM systems help keep track of prior interactions with customers, what their preferences are, what products and services they have purchased, what maintenance they have required in the past, and so forth.

cross-selling   Offering customers products or services that are complementary to the ones they are purchasing. Offering a customer to reserve a rental car or hotel room when he orders a plane ticket is a cross-selling example.

cryptographic key   See digital key.

cryptography   The science of rendering information unintelligible to all but its senders and receivers.

D-AMPS   Digital AMPS. Earlier designation of American standard IS-136, also known as TDMA.

data mining   The combing and analysis of information from one or more databases using pattern recognition technology to identify trends and patterns such as groups of customers with similar preferences.

data rate   Rate at which data is transmitted—typically measured in bits per second (bps).

decryption   The process of converting encrypted data back to its intelligible form.

digital certification   The process of establishing a digital ID for an individual, organization, or other entity by issuing a certificate.

digital key   String of bits used to scramble and unscramble messages. Given an encryption algorithm, longer keys will help achieve higher levels of security.

digital signature   A unique, unforgeable string of characters that is the electronic equivalent of a handwritten signature.

disintermediation   Process of eliminating some or all intermediaries in a business transaction or business process.

Domain Name System (DNS)   A distributed database that maps domain names to IP addresses.

dual band   Refers to telephones capable of operating in two frequency bands. In Europe, most GSM phones are dual band, operating over both the 900 and 1800 MHz GSM bands.

E112 (Europe) Enhanced GSM-based 112 emergency service that includes caller location.

E911 (USA) Enhanced 911 emergency service that includes caller location.

e-business See electronic business.

e-commerce See electronic commerce.

EDGE Originally intended as abbreviation for Enhanced Data Rates for GSM Evolution, EDGE now stands for Enhanced Data Rates for Global Evolution. EDGE is a 3G mobile communications technology based on evolution of either GSM/GPRS or TDMA network technologies. The GSM/GPRS evolution is known as Classic EDGE, while the TDMA evolution is referred to as Compact EDGE.

electronic business Any business process involving more than one organization that is electronically implemented over the Internet.

electronic commerce The purchasing and selling of goods over the Internet.

electronic signature See digital signature. EMV Standard developed by Europay, Mastercard and Visa to replace magnetic stripe credit cards with chip-based credit cards. In addition to the added simplicity and convenience associated with smartcards, EMV has been designed to support higher levels of security.

encryption A function used to translate data into a secret code, transforming its content to hide the information and, therefore, prevent its unauthorized use.

end-to-end security Ensuring security in communication and transactions from point of origin to point of destination.

E-OTD Enhanced Observed Time Difference. A handset-based positioning technology.

EPOC Mobile service operating system offered by the Symbian consortium.

e-tailer Online retailer.

e-tailing Retailing over the Internet.

ETSI European Telecommunications Standards Institute.

extranet An intranet that has been extended to include access to or from selected external organizations such as customers or suppliers.

FA Foreign agent. A router responsible for detunneling and forwarding packets it receives for a mobile node (from that mobile node's home agent).

FCC Federal Communications Commission.

FDD Frequency division duplex. Separating uplink and downlink traffic through allocation of distinct frequency bands.

FDM Frequency division multiplex. Basic multiplexing scheme used to separate channels by assigning them non-overlapping frequency bands.

gateway A node or switch that permits communications between two dissimilar networks. The WAP gateway, for example, helps translate between WAP protocols and regular fixed Internet protocols.

GGSN Gateway GPRS Support Node. Acts as an interface between the GPRS network and the Internet, giving to the outside world the impression that each mobile station operates as a regular Internet node.

GHz Gigahertz. One GHz = 1 billion hertz or 1 billion cycles per second.

GIS Geographic Information System. A combination of elements designed to store, retrieve, manipulate, and display geographic data.

General Packet Radio Services (GPRS) A 2.5G packet-switched mobile communication technology designed as enhancement of GSM networks.

Geocoding Conversion of street addresses and other human-understandable location designations into geographic coordinates (that is, latitude and longitude).

global positioning system (GPS) A U.S. government-owned technology based on the use of three or more satellites (triangulation) to provide 24-hour positioning information that indicates the precise location of any compatible receiver unit. Until May of 2000, this system was subject to the government's controlled dilution of precision (DOP), reducing accuracy to 100 meters. Without DOP, GPS now supports 5- to 40-meter accuracy.

Global system for mobile communication (GSM) Formerly known as Groupe Spécial Mobile, the world's most widely used second-generation mobile communication system. Implemented in the 900 MHz and 1800 MHz bands in Europe, Asia, and Australia, and the 1900 MHz band in the United States.

GSMA GSM Association. Consortium of over 500 network operators, infrastructure equipment manufacturers, and other technology and application providers focusing on GSM, GPRS, EDGE, and UMTS mobile communication solutions.

GSN GPRS support node (also referred to as Serving GPRS support node, or SGSN). Acts as a router that buffers and forwards packets to the mobile station via the proper BSC, as the user roams from one area to another.

GTP GPRS Tunneling Protocol. Early (and partial) implementation of mobile IP in the GPRS mobile communication standard.

HA Home agent. Router responsible in Mobile IP for intercepting packets destined to a mobile node and for tunneling them to the mobile node's care-of-address.

handover The action of switching a call in progress without interruption from one cell to another (intercell) or between radio channels within the same cell (intracell), as the caller moves out of range or the network automatically reroutes the call.

hash A number generated from a string of text as the result of a computation that turns variable-size input into a fixed-size string, used in the signing of messages. The hash is significantly smaller than the text itself, and is generated in such a way that it is extremely unlikely that some other text would produce the same value.

HDML Handheld Device Markup Language. First hypertext markup language to introduce the deck of cards metaphor, HDML is a precursor of WAP.

high-speed circuit-switched data Circuit-switched mobile communication technology that speeds up GSM transmission rates by allowing a given channel to be allocated multiple time slots in each TDMA frame. HSCSD can reach peak rates of about 40 kbps.

HLR Home Location Register. Key database in the GSM architecture used, among other things, to store the user's personal information such as his telephone number and authentication key.

HSCSD See high-speed circuit-switched data.

HTML HyperText Markup Language. The original Web markup language.

HTTP HyperText Transfer Protocol. Stateless Internet protocol that handles the transfer of Web pages. Basic HTTP functions include GET (to fetch an object from a Web server), PUT (to upload data to the server), POST, and HEAD.

hyperlink A word or phrase highlighted in a Web document that acts as a pointer to another document, whether on the same server or on a different one.

Hz (Hertz) A radio frequency measurement where one hertz equals one cycle per second.

IEEE 802.11 Currently the most popular wireless LAN standard. Original versions were designed to run in the 2.4-GHz band and support data rates between 1 and 2 Mbps. More recent enhancements of the standard have been developed for the 5-GHz range and support data rates in excess of 20 Mbps.

i-Mode NTT DoCoMo's mobile Internet service and its portal. As of late 2001, the service, which was launched in February 1999, had 30 million users.

IMT International Mobile Telecommunications.

IMT-2000 International Mobile Telecommunications 2000. An ITU initiative aimed at developing worldwide standards for third-generation mobile communication systems.

infotainment   Information and entertainment content.

Integrated Services Digital Network (ISDN)   Service that provides digital connectivity for simultaneous transmission of voice and data over multiple communication channels. ISDN supports data speeds three to four times faster than the common 56-kbps modems.

integrity   Ensuring that the content of a message or transaction is transmitted free from error, alteration, or corruption.

intelligent agent   See agent.

Internet   A worldwide collection of interconnected networks that functions as a single virtual network.

Internet service provider (ISP)   Organization that provides access to the Internet either via modem or T1 line. ISPs also have exclusive content, databases, and online discussion forums available to their users.

intranet   A private network accessible only to employees within an organization, intranets use the Internet suite of protocols to support communication, collaboration, and access to enterprise databases and other corporate resources.

IP   Internet Protocol.

ISDN   Integrated Services Digital Network.

ISP   Internet service provider.

issuer   A financial institution that issues payment cards to cardholders and guarantees payment for authorized transactions.

ITU   International Telecommunications Union. The telecommunications industry worldwide standards body.

Java   *Write Once, Run Anywhere* programming language that runs on virtual machines. The first Java-enabled mobile phones were introduced in late 2000.

J2ME   Java to Micro Editor.

JVM   Java Virtual Machine.

kbps   Kilobits per second: data transmission speed measurement unit.

key   See digital key.

key pair   A pair of digital keys—one public and one private—used for encrypting and signing digital information. A message is encoded with a private key and can only be decoded with the corresponding public key, and vice versa.

KVM   Kilo (or K) virtual machine.

LAN   See local area network.

LBS   Location-based services.

local area network (LAN)   A computer network designed for use within a building or some other relatively small geographical area (for example, airport, hospital, university campus).

LIF   Location Interoperability Forum. Industrial forum whose objective is to promote the development of location-based solutions capable of operating across networks that implement different location-tracking technologies.

location-sensitive services/applications   Services or applications capable of presenting users with information directly relevant to their current position.

location tracking   Technology used to track the location of the user as he moves from one area to another.

market share   A company's sales expressed as a percentage of the sales for the total industry.

mass customization   Mass production of customized products and/or services.

m-commerce   Mobile e-commerce.

message digest   Concisely represents a longer message or document from which it was computed. It is used to create a digital signature unique to a particular document. A message digest can be made public without revealing the contents of the document from which it is derived.

MExE   Mobile Execution Environment. A 3GPP application development environment aimed at avoiding fragmentation of the mobile application development landscape by defining three categories of devices (or *classmarks*).

microbrowser   Web browser specifically designed to access the Web from a mobile device such as a mobile phone.

MIDP   Mobile Information Device Profile. A special set of device-specific Java APIs developed for J2ME's CLDC configuration.

MIME   See Multipurpose Internet Multimedia Extension.

MMS   Multimedia Messaging Services.

mobile carrier   Same as mobile network operator.

Mobile Information Device Profile (MIDP)   A special set of device-specific Java APIs developed for J2ME's CLDC configuration.

mobile (network) operator   The wireless carrier that provides mobile communications to the end user (subscriber).

modulation   The process of encoding data for transmission over the air interface.

MP3   The file extension for MPEG audio player 3. A standard technology for the compression of sound into very small files; therefore, a popular music download format.

MPEG Moving Pictures Experts Group. An ISO group responsible for setting the standards for compression and storage of audio, video, and animation in digital form.

MPEG4 Compression standard for video telephony.

MS Mobile station.

MSC Mobile switching center.

MT Mobile terminal.

Multipurpose Internet Multimedia Extension (MIME) An Internet protocol for sending email messages and attachments.

MultOs An operating system for smartcards.

non-repudiation Ensuring that parties to a transaction cannot falsely claim later that they did not participate in that transaction.

operating system (OS) Program responsible for managing a device's many resources.

opt-in Process of actively granting an e-marketer permission to send a user a promotional message or to collect personal information for marketing purposes.

opt-out Process of notifying an e-marketer that a user does not wish to receive any promotional messages or does not wish to have his or her personal information collected for marketing purposes.

OSA Open Service Access. 3GPP initiative aimed at developing a set of APIs that third party application developers and content providers will be able to use to access key mobile network functionality (for example, charging support services, user location information, SMS applications).

P3P Platform for Privacy Preferences. W3C standard used to describe a website's privacy policy.

packet Short message of 128 bytes in length. Packets are used to measure data traffic generated by a user and compute his or her monthly bill.

packet-switched A transfer mode in which data is sent in small chunks or packets, making it possible for multiple channels to dynamically share bandwidth.

Palm OS Operating system running on PDAs manufactured by Palm and other vendors such as Handspring.

Palm Query Applications Web-clipping applications designed to query databases and access elements of Web pages from Palm OS PDAs.

PAN Personal area network. A type of wireless network designed for deployment in an individual's personal space.

PARLAY An open multi-vendor consortium working together with 3GPP and other standardization bodies on the development of open APIs to enable third-party development providers to access mobile communication network functionality across different network standards.

PCS Personal Communications Services. Term used by the FCC to designate a set of digital mobile communication services operating in the 1900-MHz frequency band in the United States.

PDA Personal digital assistant.

PDC Personal Digital Cellular. Dominant 2G cellular network standard in Japan.

PDC-P Packet-switched data network associated with PDC.

permission marketing A marketing method whereby companies get their customers' permission to market products or services to them.

personal digital assistant (PDA) A handheld computer that serves as organizer, electronic book, and note taker. Typically used with a stylus or pen-shaped device for data entry and navigation.

personalization The tailoring of services to an individual's needs or preferences.

PHS Personal Handyphone System. Japanese digital mobile communication standard operating in the 1900-MHz band.

PIM Personal information management. Includes applications such as calendars and address books.

PIN Personal identification number.

PKI Public Key Infrastructure. A security architecture that combines specialist authorities, digital certificate management systems, and directory facilities to provide security on top of unsecure networks such as the Internet.

Pocket PC Device using the third version of Microsoft's WindowsCE operating system; also the name of that operating system.

portal Major Web hub that guides users to all types of resources and services through links and functions. Resources typically include search engines, directories of services, email, calendar systems, and instant messaging.

PQA See Palm Query Applications.

privacy The control over the access to one's personal information by others as well as its use.

private key One half of a key pair, kept private by the owner and used, in conjunction with a matched public key, to sign documents and encrypt or decrypt messages or files.

profiling   Techniques used to develop or refine user characteristics.

PSE   Personal Service Environment. A common subscriber profile proposed by 3GPP for possible use across all 3G services. The profile would contain the subscriber's personal details such as his or her telephone number, email address, street address, subscription information, and payment details, as well as preferences that are likely to be re-usable across a number of different services.

PSN   Public Switched Network. Also referred to as PSTN.

PSTN   Public Switched Telephone Network. The traditional fixed analog telephone infrastructure.

public key   One-half of a key pair, used to verify signatures created with a matched private key. Also used to encrypt messages or files that can only be decrypted using the matched private key.

public key certificate   Binds a public key value to a set of information that identifies the entity, (in other words, person, organization, account, and so forth) associated with the use of the corresponding private key.   The binding, along with assurance of integrity, is provided by a CA's signature of the certificate.

public key cryptography   An encryption method that uses public and private key pairs. Information encrypted with one key can only be decrypted using the other.

Public Key Infrastructure (PKI)   A set of policies, processes, and technologies that use public key cryptography and key certification practices to secure communications.

push   A method of interaction where content is sent to the user rather than requested (*pulled*) by the user.

quality of service (QoS)   Generic term referring to a collection of techniques and parameters aimed at providing users with a certain level of consistency in the service they receive from a network.

registration authority (RA)   An independent, third-party trusted agent who verifies and approves (or rejects) pending individual certificate requests.

reverse billing   Practice where mobile operators charge users a premium for receiving SMS messages.

revocation   Process by which a certification authority revokes an otherwise valid certificate. Revoked certificates are posted to a publicly accessible Certificate Revocation List (CRL).

RF   Radio frequency.

roaming   The ability of a user to access wireless telecommunication services in areas other than his or her home subscription area.

root certificate or key　Certificate at the top of the certificate hierarchy. The root certificate, which is self-signing, is used to sign all subordinate CA certificates.

router　A data switch that handles connections between different networks. A router identifies the addresses on data passing through the switch, determines which route the transmission should take, and collects data in *packets* that are sent to their destinations.

RSA　The RSA public-key cryptosystem. Named after its developers, Rivest, Shamir, and Adleman.

Secure Electronic Transaction (SET)　A set of cryptographic protocols designed to ensure secure credit card payments over the Web through authentication of both consumers and merchants.

Secure Socket Layer (SSL)　Internet protocol designed to encrypt communication between a browser and a Web server.

Semantic Web　An ambitious initiative aimed at making Web content machine-understandable. In particular, this includes efforts to develop extensions to emerging Web Services standards in support of intelligent agent functionality.

server　A computer that manages one or more network resources. A Web server manages Web content.

service provider　An entity responsible for the provision of a service or a set of services to the user.

SET　Secure Electronic Transaction. Open technical standard developed by Visa and Matercard for online credit card payments.

SET certificate　Certificates that include SET private extensions to the X.509 certificate format and are used exclusively as a payment protocol in electronic commerce for SET cardholders, merchants, or payment gateways.

SGSN　Serving GPRS Support Node. See GSN.

signature　See electronic signature and digital signature.

SIM　Subscriber Identification Module. The smartcard helps authenticate users in the GSM/GPRS standard..

smartcard　A plastic card with an embedded chip that permits tamper-resistant storage. Besides secure storage, smartcards often feature sufficient computing power to also perform cryptographic functions.

SMS　Short Message Service. A service for sending messages of up to 160 characters to mobile phones. SMS is embedded in GSM and its 2.5G and 3G evolutions. Similar functionality is becoming available in a number of other standards, such as TDMA or CDMA.

SOAP   Simple Object Access Protocol. A lightweight XML/HTTP-based protocol for accessing services, objects, and servers in a platform-independent manner.

SMS-C   Short Message Service Center.

spamming (or **spam**)   Sending unsolicited email and messages.

SSL   See Secure Socket Layer.

stickiness   The length of time visitors remain with a site; sometimes also used to refer to a user's propensity to return to a site.

STK   SIM Tool Kit.

subscriber   A person or other entity that has a contractual relationship with a service provider and is responsible for possible access charges.

symmetric encryption   An encryption method where both communicating parties use the same key to encrypt and decrypt messages. Also referred to as *secret-key cryptography.*

synchronization   The process of replicating information between a mobile device and a fixed device.

TA   Timing Advance. A network-based positioning solution used to estimate how far the mobile station is from the base station. This technique is often used to complement CGI.

TCP/IP   Transmission Control Protocol/Internet Protocol. Also used to generically refer to the Internet suite of protocols.

TDD   See Time Division Duplex.

TDMA   See Time Division Multiple Access.

telematics   Refers to a broad collection of in-car wireless services, that can range from navigation to email and entertainment, and emergency and roadside assistance.

terminal   The equipment used by an end user to interface with a network.

(terminal) mobility   The ability of a terminal to access telecommunication services from different locations and while in motion, and the capability of the network to identify and locate that terminal or the associated user.

TIA   Telecommunications Industry Association. U.S. telecommunications standards body.

Time Division Duplex (TDD)   Separates uplink and downlink traffic through allocation of different time slots within the same frequency band.

Time Division Multiple Access (TDMA)   A type of digital cellular network in which calls are sliced into time periods and interleaved with others on the same channel.

Time To First Fix (TTFF) The time it takes to locate the satellites and compute the handset's position when using GPS.

TLS Transport Layer Security. Protocol that now replaces SSL.

TOA Time of arrival. A network-based triangulation technique that relies on uplink signals sent by the mobile station to three or more base stations to compute the handset's position.

transaction In the commercial sense, contractual agreement or exchange of goods or services possibly in return for payment. In the technological sense, the completion of complementary data updates in two or more entities.

UA User agent. Software that interprets WML, WMLScript, WTAI, and other forms of Web content. Explorer, Netscape, UP.browser, and Mobile Explorer are all examples of UAs.

UDDI Universal Description, Discovery, and Integration. An emerging standard aimed at providing a platform-independent, open framework for describing, advertising, and discovering Web services. The standard includes a registry where Web service descriptions are advertised.

Universal Mobile Telecommunications System (UMTS) UMTS/WCDMA is the 3G wideband CDMA standard jointly developed by Europe and Japan.

up selling Selling upgrades, add-ons, or enhancements to a particular product or service.

URL Uniform Resource Locator or Web address.

USIM Universal Subscriber Identity Module. UMTS evolution of the SIM module.

value chain The different entities/organizations that contribute to the creation of a particular good or service and its delivery to the customer.

VHE Virtual Home Environment. 3GPP's vision of providing users with a consistent experience independently of the network through which they connect as they roam from one operator or even one country to another.

VLR Visitor Location Register. A mobile switching center's (MSC) database that stores information about all users currently in the MSCs location area, including information obtained from each user's Home Location Register (HLR).

VPN Virtual private network. A virtual intranet consisting of multiple physical networks separated by a public network. VPNs allow corporate users to access the company's intranet even when not on company premises.

Voice XML Voice eXtensible Markup Language (formerly VXML). A markup language designed to support Web access through audio dialogues featuring synthesized speech, digitized audio, and speech recognition input.

W3C   World Wide Web Consortium.

WAA   Wireless Advertising Association.

WAE   See Wireless Application Environment.

walled garden   A collection of mobile commerce services offered by a network operator, device manufacturer, or portal to the user as the default or only set of mobile commerce options available to him or her.

wallet   Program designed to hold electronic credit cards and their associated certificates, and used in the context of the SET protocol. Mobile variations of SET have been proposed that involve placing the wallet on a server rather than on the mobile device, which today would be impractical.

WAN   Wide area network.

WAP   See Wireless Application Protocol.

WAP Forum   Industrial forum in charge of developing and promoting the WAP suite of protocols.

WAP gateway   Gateway responsible for interfacing between the WAP protocol stack and the fixed Internet protocol, and for assisting with a number of other key activities (for example, authentication, billing, push messages, and so forth).

WAP Identity Module or **Wireless Identity Module (WIM)**   A tamper-proof component designed as part of the WAP architecture to store private data (such as key pairs, certificates, and PIN numbers) within a mobile device. In practice, a WIM is implemented using a smartcard, which might or might not be the same as the one on which the SIM resides.

WAP server   A Web server delivering WML/WMLScript documents or XHTML Basic documents.

WASP   See wireless ASP.

WCDMA   Wideband Code Division Multiple Access. Essentially the same 3G standard as UMTS.

WDP   Wireless Datagram Protocol. Protocol interfacing between WAP's higher-level protocols and any one of the many bearer services supported by WAP (for example, GSM, cdmaOne, GPRS, SMS).

Web clipping application   An early mobile Internet solution introduced by Palm for its PDAs in the context of its Palm.net Web service.

Web server   A server that manages Web content accessible by outside browsers.

Web Services   Emerging set of standards aimed at providing a service-oriented, component-based architecture in which discrete tasks within e-business processes can be distributed across an arbitrary number of players. In its current form, the Web Services initiative revolves around three languages: the Simple Object Access Protocol (SOAP), Universal Description, Discovery, and Integration (UDDI), and the Web Services Description Languages (WSDL).

Web site   A collection of interconnected Web pages.

WIM   See WAP Identity Module.

WindowsCE   Microsoft operating system family for handheld and portable devices.

Wireless Application Environment (WAE)   The part of the WAP standard that includes specifications of the WAP microbrowser, the WAP markup and scripting languages, and the Wireless Telephony Applications Interface (WTAI).

Wireless Applications Protocol (WAP)   De facto wireless Internet standard capable of running on top of almost any bearer service. WAP2.0, which was released in the summer of 2001, introduced a second WAP stack specifically intended for fast bearers with built-in IP.

Wireless ASP   Wireless application service provider. Used to refer to a broad category of mobile players that operate as application service providers (ASP).

WLAN   Wireless local area network.

WLIA   Wireless Location Industry Association.

WML   Wireless Markup Language. WAP's original markup language, which, like HDML, relies on the *deck of cards* metaphor. With the introduction of WAP 2.0, WML is also sometimes referred to as WML1 in contrast to WML2, which refers to XHTML Basic. WAP 2.0 allows for content written in both WML and XHTML Basic. Note also that WML is an XML language.

World Wide Web   Global hypertext information system that is distributed throughout the world and uses the Internet suite of protocols as transport mechanism.

WSDL   Web Services Description Language. An emerging XML standard for describing how to access Web services.

WSP   Wireless Session Protocol. Provides the upper-level application layer of WAP with a consistent interface for two session services, a connection-mode service that operates above a transaction layer protocol, and a connectionless service that operates above a secure or non-secure datagram transport service.

WTLS   Wireless Transport Layer Security. The WAP equivalent to TLS (Secure Sockets Layer). The WTLS layer is designed to provide privacy, data integrity, and authentication between two communicating applications, while operating over wireless links with possibly low bandwidth and high latency.

WTP Wireless Transaction Protocol. Acts as a WAP substitute for TCP, while also handling some HTTP functionality, and has been optimized for operation over the wireless link.

X.509 certificate The most commonly used format for certificates as defined by an international standard called *X.509*. Any application complying with X.509 can read or write certificates.

XHTML eXtensible Hypertext Markup Language. A reformulation of HTML 4 as an XML standard; XHTML stabilizes the functionality of HTML, adds extensibility to HTML, and improves portability of HTML code.

XML eXtensible Markup Language. A means of defining documents with associated structure and semantics. A W3C standard for Internet Markup Languages.

References and Other Relevant Resources

In writing this book, I have drawn on a variety of sources of information, from news articles, conversations, and emails with a number of people, to online documents, scientific publications, and various books. The following is a list of publicly available online resources, scientific articles, research reports, books, and other publications that I have found to be particularly useful. Clearly, in a fast-moving area such as mobile commerce, I cannot guarantee that all the online resources listed here will continue to exist, let alone reside at the same URLs. Information about teaching material such as slides or videotapes of my lectures will also be made available on my Carnegie Mellon University Web site (www-2.cs.cmu.edu/~sadeh/).

Books, Reports, and Scientific Publications

Andersson, Christoffer, *GPRS and 3G Wireless Applications*, John Wiley & Sons, ISBN 0-471-41405-0, 2001.

Barnett, Nick, Stephen Hodges, and Michael J. Wilshire, "m-Commerce: An Operator's Manual," *The McKinsey Quarterly*, Number 3, pp. 163–173, 2000.

Bergman, Eric, and Robert Haitani, "Designing the Palm Pilot: A Conversation with Rob Haitani," Chapter 4, in *Information Appliances and Beyond*, edited by Eric Bergman, Morgan Kaufman Publishers, ISBN 1-558-60600-9, 2000.

Berners-Lee, T., J. Hendler, and O. Lassila, "The Semantic Web," Scientific American, May 2001.

Bughin, Jacques R., Fredrik Lind, Per Stenius, and Michael J. Wilshire, Mobile Portals Mobilize for Scale, *The McKenzie Quarterly*, Number 2, pp.118–127, 2001.

Burke, R., "The Wasabi Personal Shopper: A Case-Based Recommender System," in Proceedings of the 16th National Conference on Artificial Intelligence (AAAI-99), American Association for Artificial Intelligence, 1999

Christensen, Erik, Francisco Curbera, Greg Meredith, and Sanjiva Weerawarana, "Web Services Description Language (WSDL) 1.1."

Cotter, Paul, and Barry Smyth, "WAPing the Web: Content Personalization for WAP-enabled Devices," pp. 98–108, in "Adaptive Hypermedia and Adaptive Web-based Systems," AH2000 Conference Proceedings, Eds. P. Brusilovsky, O. Stock, and C. Strattarava, *Lecture Notes in Computer Science 1892*, Springer Verlag, ISBN 3-540-67910-3, 2000.

Cunningham, P., R. Bergmann, S. Schmitt, R. Traphöner, S. Breen and B. Smyth, "WEBSELL: Intelligent Sales Assistants for the World Wide Web," Trinity College Technical Report TCD-CS-2000-42, Trinity College, Ireland, November 2000

The DAML Services Coalition, "DAML-S: Semantic Markup for Web Services," DAML-S (www.daml.org), July 2001.

Daum, Adam, "Selling m-Commerce in Europe: No Killer Application Just Killer Attitude," Gartner report, March 2001.

Devine Alice, and Sanna Holmqvist, "Mobile Internet Content Providers and Their Business Models—What Can Sweden Learn from the Japanese Experience?" Master Dissertation, Department of Industrial Engineering and Management, The Royal Institute of Technology, Stockholm, Sweden, 2001.

Durlacher Research, Mobile Commerce Report, November 1999. Can be downloaded at www.durlacher.com/fr-research.htm.

Eriksson, C., and F. Norin, "Female Applications for the m-Generation," NetLight Consulting AB, 2000.

European Information Technology Observatory, "Mobile e-Commerce: Market Perspectives," pp. 256–281, EITO 2001, 2001.

Fridman, Jonas, "Billing in Change, a Peek into Billing in 3G Environments," edgecom, June 2001.

Halsall, Fred, *Data Communications, Computer Networks and Open Systems, Fourth Edition*, Addison Wesley, ISBN 0-201-42293-X, 1996.

Hatloy, Andres S., "Strategies and Scenarios for Wireless Information Services," Master of Science in the Management of Technology Dissertation, MIT, June 2000.

Hayward, Chris, "The Outlook for mCommerce: Technologies and Applications to 2005," Reuters Business Insight, October 2000.

Healey, Nick, "The EPOC User Interface in the Psion Series 5," Chapter 6, in *Information Appliances and Beyond*, edited by Eric Bergman, Morgan Kaufman Publishers, 2000.

Hjelm, Johan, *Designing Wireless Information Services*, John Wiley & Sons, ISBN 0-471-38015-6, 2000.

Hjelm, Johan, *Creating Location Services for the Wireless Web: Professional Developer's Guide*, John Wiley & Sons, ISBN 0-471-40261-3, 2002.

Hoffman, Matthew, and Matthew Lewis, "The Impact of GPRS," Wit SoundsView Corp., 2001.

Information Society Technologies Advisory Group (ISTAG), "Scenarios for Ambient Intelligence in 2010," Report compiled by K. Ducatel, M. Bogdanowicz, F. Scapolo, J. Leijten, and J-C Burgelman, Office for Official Publications of the European Communities, ISBN 92-894-0735-2, 2001.

Kronenberg, Abi, "Keep billing simple for your customer," edgecom, June 2001.

Krueger, Malte, "The Future of m-Payments: Business Options and Policy Issues," Electronic Payment Systems Observatory, Institute for Prospective Technological Studies, Joint Research Center, European Commission, Report EUR 19934 EN, August 2001. Can be downloaded at epso.jrc.es/Docs/Backgrnd-2.pdf.

Little, Arthur D., "Serving the Mobile Customer, 2000." Can be downloaded at www.adlittle.com/management/services/telecom/articles/e-Mobility.pdf.

Macskassy, S.A., A.A. Dayanik and H. Hirsh, "EmailVallet: Learning User Preferences for Wireless Email," Proceedings of the IJCAI-99 workshop on "Learning About Users and Marchine Learning for Information Filtering," Sweden, 1999

McGuire, Mike, "The m-Commerce Promise: One Part Reality, Three Parts Dream," Gartner report, May 2001.

McIlraith, S. A., T. C. Son, and Honglei Zeng, "Semantic Web Services," IEEE Intelligent Systems, March/April 2001.

Narayan, D., Flinn, J., Satyanarayan, M., "Using History to Improve Mobile Application Adaptation," Proceedings of the Third Workshop on Mobile Computing Systems and Applications, Monterey, CA, Dec. 2000

Nokia, "The Demand for Mobile Value-Added Services: Study of Smart Messaging," 1999.

Norman, Donald, *The Invisible Computer*, MIT Press, ISBN 0-262-14065-9, 1998.

Odlyzko, A., "The Visible Problems of the Invisible Computer: A Skeptical Look at Information Appliances," www.firstmonday.org/issue4_9/odlyzko/.

Pesonen, Lauri, "GSM Interception," Department of Computer Science and Engineering, Helsinki University of Technology, November 1999.

Rader, Michael, and Ulrich Riehm, "Mobile Phone Payment Systems," *Electronic Payment Systems Observatory Newsletter*, No. 2, October 2000. Special issue on mobile payments. Can be downloaded at epso.jrc.es/newsletter.

Ramasundaram, Sethuraman, "WSDL: An Insight Out," *XML Magazine*, pp. 40–44, Vol. 2, Number 5, October/November 2001.

Ramsay, Marc and Jakob Nielsen, "WAP Usability Déjà Vu: 1994 All Over Again—Report from a Field Study in London, fall 2000," *Nielsen Norman Group report*, December 2000.

Sadeh, Norman, "A Semantic Web Environment for Context-Aware Mobile Services," Wireless World Research Forum Conference, Stockholm, September 2001.

Schiller, Jochen, *Mobile Communications*, Addison Wesley, ISBN 0-201-39836-2, 2000.

Shardanand, U. and P. Maes, "Social Information Filtering: Algorithms for Automating 'Word of Mouth," in Proceedings of the 1995 Conference on Human Factors and Computing Systems (CHI'95), Denver, CO, 1995 (ISBN 0-89791-694-8)

Shearin, S. and H. Lieberman, "Intelligent Profiling by Example," in Proceedings of the International Conference on Intelligent User Interfaces (IUI2001), pp. 145-152, Santa Fe, NM January 2001.

Sherif, Mostafa Hashem, *Protocols for Secure Electronic Commerce*, CRC Press, 0-8493-9597-6, 2000.

Sierra, C., M. Wooldridge, and N. Sadeh, "Agent Research and Development in Europe," IEEE Internet Computing, September/October 2000.

Skvaria, Carol, and Brian Dooley, "Wireless Services: United States," Gartner report, November 2001.

Small, J., Smailagic, A., and Siewiorek, D., "Determining User Location For Context Aware Computing Through the Use of a Wireless LAN Infrastructure," December 2000.

Solomon, James D., *Mobile IP: The Internet Unplugged*, Prentice Hall, ISBN 0-13-856246-6, 1998.

Sousa, J.P., and Garlan, D., "Aura: An Architectural Framework for User Mobility in Ubiquitous Computing Environments," Proceedings of the Third Working IEEE/IFIP Conference on Software Architecture, Montreal, August 2002.

Steels, Elizabeth, *Creating and Sharing Value in the Mobile Ecosystem— Choosing and Intelligent Business Model for Content and Messaging Applications*, OpenWave white paper, September 2001.

Stone, Gavin, "MexE: Mobile Execution Environment," MexE Forum, December 2000.

Tainio, Antti (Nordea), "Wired Banking Goes Wireless," Private Communication, June 2001.

Tanenbaum, Andrew S., *Computer Networks, Third Edition*, Prentice Hall, ISBN 0-13-349945-6, 1996.

Tober, Eric, and Robert Marchand, "VoiceXML Tutorials," Available at www.voicexml.org/tutorials, 2001.

Treese, G. Winfield, and Lawrence C. Stewart, *Designing Systems for Internet Commerce*, Addison Wesley, ISBN 0-201-57167-6, 1998.

Turban, Efraim, Jae Lee, David King, and H. Michael Chung, *Electronic Commerce: A Managerial Perspective*, Prentice Hall, ISBN 0-13-975285-4, 1999.

UMTS Forum, Reports 1–16, 1997–2001. Can be downloaded at www.umts-forum.org.

Väänänen-Vainio-Mattila, Kaisa, and Satu Ruuska, "Designing Mobile Phones and Communicators for Comsumers' Needs at Nokia," Chapter 7, in *Information Appliances and Beyond*, edited by Eric Bergman, Morgan Kaufman Publishers, ISBN 1-558-60600-9, 2000.

WAP Forum, WAP 2.0 Technical White Paper, August 2001. Available at www.wapforum.org.

Wieland, Ken, "Where are the location-based services?" *Telecommunications Online, International Edition*, September 2000. Can be downloaded at www.telecoms-mag.com/telecom.

Windwire, First-To-Wireless Report, Windwire Inc, Morrisville, NC 27560, 2000. Available at www.windwire.com.

Youll, J. J. Morris, R. Krikorian, and P. Maes, "Impulse: Location-based Agent Assistance," Proceedings of the Fourth International Conference on Autonomous Agents (Agents2000), Barcelona, Spain, June 2000.

Zobel, Rosalie, and Norman Sadeh, "eBusiness and eWork: The Challenges Ahead," eBusiness and eWork 2001, Venice, October 2001.

Zuberec, Sarah, "Interaction Design and Usability of MS Windows CE," Chapter 5, in *Information Appliances and Beyond*, edited by Eric Bergman, Morgan Kaufman Publishers, ISBN 1-558-60600-9, 2000.

General M-Commerce-Related Sites

AllNetDevices (www.allnetdevices.com). General news site covering a broad range of wireless issues.

FierceWireless (www.fiercewireless.com). General wireless and mobile Internet news, including a free weekly newsletter.

Mbusiness (www.mbizcentral.com). Both a Web site with daily news and a free monthly magazine.

M-Commerce Times (www.mcommercetimes.com). A regularly updated collection of m-commerce articles.

Mobic.com (www.mobic.com). General wireless news site.

PaloWireless Wireless Resource Centre (www.palowireless.com). Another comprehensive site with reviews of technologies, applications, and business models.

Tech Buddha Wireless Obsessions (www.techbuddha.com/wireless.shtml). Wireless news site focusing on Asia.

Tech Web's Mobile and Wireless Section (www.techweb.com/tech/mobile). A mobile community portal combining news, general information, and reviews.

ThinkMobile (www.thinkmobile.com). Another mobile community portal combining news, general information, and reviews.

Wireless NewsFactor (www.wirelessnewsfactor.com). Community portal focusing on mobile and wireless news.

WirelessWeek (www.wielessweek.com). General wireless news site.

Industrial Forums, Regulatory Organizations, and Other Relevant Initiatives and Organizations

3GPP (www.3gpp.org). Third Generation Partnership Project in charge of developing specifications for the EDGE and WCDMA/UMTS standards. 3GPP organizational partners include ARIB (Japan), CWTS (China), ETSI (Europe), T1 (U.S.), TTA (Korea), and TTC (Japan).

3GPP2 (www.3gpp2.org). Third Generation Partnership Project in charge of developing specifications for the CDMA2000 standards. 3GPP2 organizational partners include ARIB (Japan), CWTS (China), TIA (North America), TTA (Korea), and TTC (Japan).

American National Standards Institute. ANSI (www.ansi.org).

Association of Radio Industries and Businesses. ARIB (www.arib.or.jp).

Bluetooth SIG (www.bluetooth.org and www.bluetooth.com). Consortium of companies focusing on the development and adoption of Bluetooth technologies.

DARPA Agent Markup Language (DAML) initiative (www.daml.org). DARPA Semantic Web research initiative, which includes development of the DAML+OIL language and the DAML-S Web service description language.

ePayment Systems Observatory—ePSO (epso.jrc.es). European ePayment Observatory at the Institute for Prospective Technological Studies of the European Commission's Joint Research Center.

European Telecommunications Standards Institute—ETSI (www .etsi.org). European telecommunications standardization organization.

Federal Communications Commission—FCC (www.fcc.org). U.S. Telecommunications Regulation Agency.

Global Mobile Commerce Interoperability Group—GMCIG (www.gmcig.org). Industry forum led by MasterCard and focusing on the development of mobile payment standards that build on related efforts such as SET and EMV.

GSM Association—GSMA (www.gsmworld.com). Wireless industry association promoting the development, adoption, and interoperability of solutions based on GSM and evolution standards such as GPRS and EDGE.

Internet Engineering Task Force—IETF (www.ietf.org). A large international community of network designers, operators, vendors, and researchers concerned with the evolution of the Internet architecture and its many protocols. Work of the IETF is organized around working groups focusing on particular protocols or issues, and has included the development of key protocols such as IP, TCP, UDP, and so forth.

International Telecommunications Union—ITU (www.itu.int). International organization coordinating global telecom network and services activities with governments and the private sector.

Location Interoperability Forum—LIF (www.locationforum.org). Industry forum formed by Ericsson, Motorola, and Nokia in September 2000 to promote the development of interoperable solutions for mobile location-based services.

MexE Forum (www.mexeforum.org). Group of companies involved in promoting, implementing, and developing applications for 3GPP's MexE specifications.

Mobey Forum (www.mobey.org). Industry forum created to promote the development of interoperable protocols for mobile financial services, including mobile payment.

Mobile electronic Transactions—MeT (www.mobiletransaction .org). Industry initiative founded in April 2000 by Ericsson, Nokia, and Motorola to promote the development of new mobile payment protocols, building on WAP, PKI, and Bluetooth technologies.

Mobile Entertainment Forum—MEF (www.mobileentertain- mentforum.org). Forum created in February 2001 by a group of mobile entertainment companies, mobile operators, and mobile technology providers to promote the development of standard revenue sharing practices and standards for cross-operator delivery of mobile entertainment services.

Mobile Gaming Interoperability Forum—MGIF (www.mgif.org). Forum founded in July 2001 by Ericsson, Motorola, Nokia, and Siemens to define mobile games interoperability specifications and application programming interfaces.

Mobile Marketing Association—MMA (www.waaglobal.org). Organization created in January 2002 through the merger of the Wireless Advertising Association (WAA) and the Wireless Marketing Association (WMA). MMA focuses on the development of wireless advertising standards and practices.

Mobile Payment Forum (www.mobilepaymentforum.org). Mobile payment forum founded by MasterCard, Visa International, American Express, and JCB in November 2001.

mSign (www.msign.org). Mobile Electronic Signature consortium founded in December 1999 by Brokat to define a worldwide mobile digital signature standard. mSign merged with Radicchio in 2001. As of early 2001, mSign had 35 partners.

Radicchio (www.radicchio.org). An industry initiative founded in the fall of 1999 by Sonera SmartTrust, Gemplus, EDS, and Ericcson whose mission is to promote the development and adoption of wireless PKI solutions.

Telecommunications Industry Association—TIA (www.tiaonline.org). North American industrial telecommunications association with over 1,100 member organizations.

Telecommunications Technology Association—TTA (www.tta.or.kr/english/e_index.htm). Industrial telecommunications association of Korea.

Universal Description, Discovery and Integration—UDDI (www.uddi.org). Standardization initiative aimed at providing mechanisms to advertise and discover Web services.

UMTS Forum (www.umts-forum.org). Forum whose objective is to promote the launch of the 3G UMTS standards. As of late 2001, the UMTS forum had over 250 members, including a number of mobile operators, infrastructure and technology platform providers, content providers, and regulatory organizations from over 40 countries.

VoiceXML Forum (www.voicexml.org). VoiceXML forum created by Motorola, AT&T, and Lucent in March 1999 and now involving around 550 member organizations. The VoiceXML markup language has now been adopted by W3C with the release of VoiceXML2.0 in October 2001.

WAP Forum (www.wapforum.org). Forum founded in 1997 by Phone.com (now OpenWave), Ericsson, Nokia, and Motorola and now consisting of over 600 members cutting across all tiers of the m-commerce value chain. The WAP Forum's primary activities have centered on the development and promotion of the WAP protocol suite, a now de facto mobile Internet standard.

Wireless World Research Forum (www.wireless-world-research.org). Forum consisting of leading companies and universities whose objective is to set strategic research directions in the field of mobile and wireless communications.

World Wide Web Consortium—W3C (www.w3c.org). With over 500 member organizations, W3C, created in 1994, has been leading the development of protocols aimed at promoting Internet interoperability. W3C activities have included the development, standardization, and promotion of languages such as XML, XML Schema, XML signatures, RDF, VoiceXML, P3P, or SMIL.

Selection of European M-Commerce Research Projects

6WINIT (www.6winit.org). Development of solutions that combine the IPv6 and GPRS/UMTS standards.

ADVICE (www.advice.iao.fhg.de). Development of an interactive sales assistant to support both fixed and mobile commerce scenarios.

e-Parking. Development of a mobile solution to book and pay for parking space.

Meteore 2000 (www.meteore2000.net). Mobile micropayment solutions.

Mobicom (www.eltrun.aueb.gr/projects/mobicom.htm). Analysis of mobile commerce technologies, business models, and policy challenges.

MyGrocer (www.eltrun.aueb.gr/mygrocer). Development of mobile technologies and scenarios for the grocery industry, building on WAP and GPRS standards.

Odin. Development of mobile location-based services.

Secrets (laplace.intrasoft-intl.com/secrets). Mobile security project.

Tellmaris (www.tellmaris.com). 3D interactive maps for tourists and people on the move.

Verificard (www.verificard.com). Development platform supporting the formal specification and verification of Javacard programs.

Ward-in-hand (www.wardinhand.org). Mobile workflow management tool for doctors and nurses at hospitals.

Youngster (www.ist-youngster.org). Development of mobile context-aware technologies and services for teenagers and young adults.

N orman M. Sadeh is an associate professor at Carnegie Mellon University (CMU), where he is affiliated with the School of Computer Science, the eCommerce Institute, and the Institute for the Study of Information Technology and Society. He currently teaches, consults, and conducts research in mobile commerce, supply chain management, agent technologies, and the Semantic Web, and is also interested in the broader business, social, and policy implications associated with the emerging Information Society.

Norman recently returned to CMU from the European Commission in Brussels, where he spent five years as program manager. At the Commission, he most recently served as chief scientist (Scientific and Work Program Coordinator) of the Euro550 million (U.S. $500 million) European research initiative in "New Methods of Work and eCommerce." As such, he was responsible for shaping European research priorities in areas such as e-commerce, m-commerce, virtual enterprises, knowledge management, agent technologies, and the Semantic Web. As of December 2000, these activities had resulted in the launch of over 200 R&D projects, typically ranging between $2 million and $4 million, and collectively involving over 1,000 European organizations in both industry and academia.

Norman has been on the faculty at CMU since 1991. Prior to joining the European Commission, he co-founded and co-directed CMU's Intelligent Coordination and Logistics Lab., which he helped position as one of the premier research organizations in intelligent planning, scheduling, and e-supply chain management. There he pioneered the development, deployment, and commercialization of several novel technologies and applications through close collaboration with organizations such as IBM, Raytheon, Mitsubishi, Komatsu, the U.S. Army, Carnegie Group (now part of Logica), and NEC.

Norman received his Ph.D. in computer science at Carnegie Mellon University. He holds a master's degree in computer science from the University of Southern California, and a BS/MS degree in electrical engineering and applied physics from Brussels Free University. He is also an APICS Certified Fellow, a Fellow of the Belgian American Educational Foundation, and a member of the ACM, AAAI, and INFORMS. He has authored approximately 80 scientific publications and serves on the editorial board of several journals, including *Autonomous Agents and Multi-Agent Systems* (AAMAS) and *Electronic Commerce Research Applications* (ECRA).